When the Stars Began to Speak
A Speculative Essay and Science Fiction Story about the Possible Impact of the Success of SETI

by Steve Geller
2014

When the Stars Began to Speak
Copyright © 2014 by Stephen P. Geller. All rights reserved.
No part of this book may be used or reproduced in any manner whatsoever without written permission, except in the case of brief quotations embedded in critical articles and reviews.

ISBN-13: 978-1501038990

ISBN-10: 1501038990

by Steve Geller

Dedication: to SETI researchers past and present, who have kept up the faith and continued the search.

Table of Contents

Contents

Table of Contents	4
Preface	10
A Short History of SETI	11
SETI at Home	12
Terms and Technical Issues	13
Some Caveats and Qualifiers	16
The SETI Café	18
ET Music; Channeling ET	19
Ambience of the SETI Café	21
Starman and Friends	24
A SETI Scientist	27
Kepler Instrument Finds a Plenitude of Planets	32
The Drake Equation	33
The Fermi Paradox	35
Felix	38
Listening to ET Music	44
Music from the Void	47
Other Prospects for Contacting an ET	51
Starsong	52
ET and Religion	53
A Philosophical Doubter	62
Cosmo	66
Dr. Stern	75
Messages Arrive from HD85	79
A Conceptual Breakthrough	80
From Bits to Bytes	87
HD85 in the News	88
Orange Juice from an Orange Star	91
Dot-Dash	93
Morse from another Source	97
HD28	98
Henry Draper Catalog	102
A Conference on ET Messages	103
News Coverage of the Message Conference	110
The Dangers of Anticipatory Naming	114
Comments from a Conservative	115
A Skeptic	116

by Steve Geller

Trying to Tune In ET Radio and TV	120
Fraudsters Found	124
Music from the Void Becomes Better	126
A Signal from Cygni	132
The Cygni Announcer	134
A Spanish Voice from Epsilon Eridani	136
Common Referents	139
Talk or Music?	140
More News Reports	141
Amateurs Transcribe the Cygni Announcer	145
An Epsiloni Tries to Speak Spanish	147
Starman the Zombie	149
Mormons not surprised by ET	154
Visions from the Void	155
Beings on Big Planets	160
Are there ETs Close to Home?	166
Sending a Message to ET	170
Visiting the SSE Lab	174
A Summary of the Situation	178
Do the ETs have a Meter Stick?	181
Who Are We Hearing From So-Far?	183
What About MV?	184
Conflicts and Rebukes	186
Conservative Casts Doubts	188
Still Thinking about ET TV	189
Another Conference	191
Amateur Linguists and LEX	196
An Historian's Perspective	198
Shamisen Check	203
Still Trying to Understand an ET Language	204
Felix Group Listens to Red Dwarfs	206
What is the Gliese Catalogue?	212
Brown Dwarfs	213
Another Fraudster Found	216
A Signal from Sirius?	217
Checking on Canopus	218
Dismissing Deneb	219
Vetting Vega	219
Follow-up with Fomalhaut	220
Images from an ET	222
Serious Speculation on the MV	227
A Possible Source for MV	231
Bibliography	233

When the Stars Began to Speak

by Steve Geller

Acknowledgements
Thanks to everyone who proof-read and criticized this writing

by Steve Geller

Preface

"When the Stars Began to Speak" is a realistic science fiction story, a mixture of non-fiction and fiction, of fascinating scientific fact and entertaining speculative fun.

The story tells what might happen if, as the result of SETI using new improved technology, many ET messages suddenly are received.

I imagine myself as a science journalist who came to Berkeley to write about SETI progress. I find myself in the front row of events when the SETI research world suddenly changed from hopeful searching and never detecting anything reliable or consistent, to dealing with many mysterious ET messages from the stars.

The locale of this fictional story is Berkeley, California, home of a world-class university. Berkeley is also home to many unconventional thinkers.

Most of the action takes place in a downtown café, where the customers represent a wide range of human experiences and attitudes. The story shows how scientists, politicians, news reporters, students, religious leaders and other people might react to the reality of encountering signals from many technological civilizations, living out among the stars.

by Steve Geller

A Short History of SETI

SETI -- the Search for Extra-Terrestrial Intelligence has been going on for a long time. SETI is the collective name for research projects involving a search for ET radio messages, mostly using radio telescopes.

This also includes Optical SETI, watching for flashes of laser light from an ET.

The main US SETI project is the SETI Institute in Mountain View, CA.

Thanks to the Kepler mission, we now know that plenty of extrasolar planets exist, and that many of them might be almost Earth-like. So-far, SETI has not received radio or TV from any technical civilizations other than those here on Earth. There have been a few interesting signals, but nothing that consistently repeats, or has been successfully analyzed to show intelligent origin.

Detecting an ET signal is just the beginning. Analyzing it for intelligent content is the big job.

We don't know what frequencies ET may be using, or whether ET uses any of the same transmission and modulation methods (e.g. AM, FM, PCM, FSK, Spread-Spectrum) that we use here on Earth.

The ET technologists may well have come up with some very different ways to send electromagnetic signals – perhaps they discovered something that we on Earth have not yet thought of and don't recognize.

But the SETI researchers are hopeful that, regardless of the transmission method, some kind of intelligent pattern will be evident in the signal if intelligent beings sent it.

So SETI soldiers on, in spite of the fact that (with perhaps one exception) nothing like an interstellar communication signal has ever been received.

SETI at Home

A Berkeley-based SETI project uses a "piggyback" antenna on the big radio telescope at Arecibo, Puerto Rico. While the telescope goes about its regular radio astronomy business, the piggyback antenna follows along, pointing to wherever the main antenna is looking and capturing a separate signal which is recorded and later analyzed for narrow-band signals from space.

Even amateurs can participate in SETI research. Anyone with a fast Internet connection and a modern home computer running Microsoft Windows, Apple Mac OS X or Linux can participate in **SETI@home** by going to the UC Berkeley website http://setiathome.berkeley.edu/ and downloading the free software which facilitates using the Internet to run SETI signal analysis at home.

by Steve Geller

Terms and Technical Issues

Radio frequencies are measured in **hertz**, which is just a name for cycles per second. High frequencies are in **megahertz (MHz)**, which is millions of cycles per second.

Narrow-band signals are radio emissions that extend over only a small part of the radio spectrum, a small range in hertz.

If you tune your radio between stations, you hear hissy scratchy random noise. This is sometimes called static, because one source for such noise is local static electricity; static noise is broad-band , appearing over a wide range of the radio spectrum.

When you tune in the carrier frequency of a broadcast radio station, you have found a narrow-band signal.

Narrow-band signals are the mark of a purposely built transmitter. Natural cosmic radio sources, such as pulsars, quasars, and interstellar gas, do not make narrow-band radio signals. The signal from these objects is spread across a band of frequencies.

Since narrow-band signals are not known to occur naturally, detecting one of them could be evidence of extraterrestrial technology. The SETI researchers hope that an ET civilization might be deliberately beaming messages at Earth, or more likely, that such a civilization may be involved in sufficiently energetic electromagnetic activity that some of it leaks out into space.

A radio signal may be converted to sound by pulling out a low-frequency audio signal which modulates a much higher carrier frequency. The audio signal has been imposed in some way on the high frequency carrier. Broadcast radio stations on Earth use either Amplitude Modulation (AM) or Frequency Modulation (FM).

For a TV signal, demodulating is more complicated; there is both picture and sound. The lines of a picture must be picked out, brightness andcolor information extracted and a change of frame recognized. None of this can be done easily by guesswork on a signal of unknown origin. The de-modulator designer has to know what conventions were used to modulate the carrier. When an ET civilization is doing the modulating, it may use very different TV conventions from those we have developed on Earth.

There are many things that can interfere with catching an ET message. A radio signal received after transmission over a long distance suffers from **attenuation** – reduced signal strength due to the familiar inverse-square law, the one that causes light to dim with distance.

Signals sent through interstellar space also suffer from **interference** and **distortion**

Interference means competition from other signals, especially those in the same wavelength range. If any signal-sending ET desires to be heard, he should pick a frequency range that's relatively free of interference, but still convenient to recognize and detect.

Distortion is more subtle. Interstellar magnetic fields can bend waves. Electric fields can rotate the polarization of waves.

Signals can experience **smearing** over time, meaning that different frequency components of the signal travel at different rates and arrive at different times. This smearing problem arises with long-haul fiber optic cables.

The relative motion between the receiver and the radio source can shift the observed wavelength. This **Doppler Effect** is useful for measuring radial motion, that is, motion along the line of sight. Essentially every signal from space shows some Doppler, because every cosmic source is in motion relative to Earth.

All of these effects can be recognized, and to some extent corrected for, but overall, they limit how well SETI will pick up extraterrestrial signals.

It is possible to tell which atomic elements make up a star by looking at its light **spectrum,** which appears when light from the star is spread out by a prism to show a rainbow of colors. Emission and absorption caused by each element makes a pattern of bright and dark lines in the spectrum. The combined pattern of the lines from all the elements which are present is used to identify the spectral class of a star.

Some Caveats and Qualifiers

Before beginning this story, I need to have six things clearly understood:

ET -- The term "ET" is simply an abbreviation for "extraterrestrial," generally meaning anything not on planet Earth. In SETI discussions, ET usually means a non-human civilization of intelligent beings, residing on a planet in orbit about some star, many light-years distant from Earth. In the story, ET will either refer to one of these extraterrestrial civilizations or generically to an individual member of an ET civilization. Anyway, as you read this story, please keep in mind that in the story context, ET is not the title of a movie; it is not the nickname of any big-eyed beings alleged to visit Earth aboard a UFO.

Stars -- The astronomical locations and physical properties of the stars described in this story, for example Epsilon Eridani, Tau Ceti, 61 Cygni, HD 85512 and Gliese 581, are correct, according to published information available at the time of writing (August 2014). However, some of the emissions these named stars are said to produce in this story are fictional inventions. The ET messages specifically are entirely products of my imagination.

SETI Café -- While Berkeley, California is a more or less real place, the Berkeley restaurant called the "SETI Café" is not real, not right now anyway. But restaurants proliferate in Berkeley. It could be, perhaps as a result of reading this story, that someone will open a real SETI Café in Berkeley or somewhere else in the San Francisco Bay Area.

by Steve Geller

UCB Research -- Most of the research activities described in this story as going on at the University of California, Berkeley, such as ET music processing and ET language analysis, are fictional inventions. Specifically, the technology of Statistical Signal Enhancement (SSE) has yet to be invented – if it's even possible. Of course the SSE lab on the UC Berkeley campus is fictional. However, SETI@home is quite real; you should give it a try. The LEX language-learning clubs are real too.

People -- The characters I describe in the story are all fictional, in the sense that while my descriptions are probably inspired by encounters with real people, the characters in the story are not real people — even the ones you are sure you recognize.

Jovian Whales -- While there has been speculation by Sagan, Salpeter and others that gas bag life forms may be floating in the atmosphere of Jupiter and other gas giant planets, nobody has yet detected any signals from this alleged life.

Finally, full-disclosure on author background: I live in Berkeley. I've had a long-term amateur interest in astronomy and space travel. I'm not a scientist. I was a software engineer. I'm retired from the UC Berkeley Space Science Lab, where I developed software to control experiments aboard satellites.

Something to keep in mind: the way things traditionally go in science and fiction, nearly all of the fiction in this story will turn out to be either fact or foolishness in the future.

That completes the prologue. Now begins the story. Here is what happened **when the stars began to speak**.

The SETI Café

The SETI Café was a restaurant of about 16 tables, located on the ground floor of a building on one of the streets that branch off Shattuck Avenue in downtown Berkeley. The restaurant had astronomical and space décor and claimed to serve food that was "out of this world." Items on the menu were said to have been inspired by stars with planets, where ET beings just possibly might live. The dishes were all vegetarian.

The SETI Café was open every day, from Noon to 9 p.m.

by Steve Geller

ET Music; Channeling ET

As you walked on the sidewalk past the SETI Café just after noon, you might have heard "ET Music" coming from the restaurant speaker system. It consisted of humming, drumming, musical tones, plunks, buzzes and chirps.

These sounds were extraterrestrial. They came from stars. Radio telescopes all over the world provided signal data from when their telescope antenna was pointed toward a particular star in the sky. This signal data was received at a lab on the UC Berkeley campus, where the radio signal was processed into an audio signal called "ET Music."

The ET Music was made available as sound files on an Internet website, which Café patrons could access with their laptop computer, using the Café's free WiFi connection. In order not to interfere with each other, each customer listened to the ET Music on his own earphones.

To attract customers, especially around lunch time, the restaurant sometimes played selected ET Music on the speaker system. This is what you heard as you passed in the street.

It may or may not have been music, but it sure was extraterrestrial. ET Music originated from stars with astronomical names such as Tau Ceti, Epsilon Eridani and 61 Cygni, which are between 10 and 20 light-years distant from Earth. These stars are known to have planets, but the planets are not all necessarily Earth-like.

Some of the people who listened to the ET music claimed to get visions of faraway star systems and life on planets orbiting around them. Some listeners thought they were in actual contact with an ET being. They claimed to absorb alien culture, including their food preferences. This was called "channeling ET." Many items on the restaurant menu were inspired by channeling ET from a particular star.

Of course, some people scoffed at channeling ET, calling it new-age nonsense. The restaurant management didn't make any claims. The ET Music was offered only as entertainment for the Café customers.

Google "music of the stars" for example star sounds on the Internet, similar those that were heard at the SETI Café.

by Steve Geller

Ambience of the SETI Café

I came to Berkeley, planning to stay for a week or so to write an article about recent activity in SETI research.

On my first visit to the SETI Café, I entered about 1:00 p.m. Most of the tables were filled.

I looked around. The walls were covered with pictures of star fields and artistic representations of scenes at hypothetical planets. One picture was labeled "Epsilon Eridani." It showed a bright Sun-like star in the far distance, and in the foreground, a planet with a banded atmosphere of colored clouds. Also circling the star was a string of small rocky bodies – a belt of asteroids. This was an artist's conception of the star system, not a photograph.

I picked up a menu and sat down at a table. The fare looked interesting and possibly tasty. Out of this world? Well, the item descriptions sure were astronomical.

SETI Café Menu

Epsilon Eridani Broccoli & Nut Salad
Chopped broccoli, mixed with nuts, carrots and beans in a pleasant soy-based sauce.
Inspired by the Epsilon Eridani star system, a Sun-like star 10.4 light-years distant in the constellation Eridanus. Numerous science fiction stories feature an ET civilization at Epsilon Eridani.

Tau Tofu
Tofu chunks with various dips for flavoring.
Inspired by the Tau Ceti star system, a yellow-orange star 12 light-years distant in the constellation Cetus. Tau Ceti is another popular locale in science fiction for an ET civilization.

Betelgeuse Beet Treat
Mixture of cooked beets - red beets, yellow beets, purple beets – in olive oil.
Inspired by the star Betelgeuse, 640 light-years distant. This is the big red star at the upper left of the man-figure of the constellation Orion. Betelgeuse is one of the largest known stars. If Betelgeuse replaced the Sun in our solar system, the outer layers of Betelgeuse would go out beyond the orbit of Mars.
(Note: If you order the Betelgeuse Beet Treat, you must be able to pronounce the star name -- BET/tll/jewz).

Sirius Sustenance
Cooked blue corn and black rice,
Inspired by the famous blue-white star Sirius, 9 light-years distant in the constellation Canis Major.

by Steve Geller

Vulcan Krei'la
A wholesome and satisfying flatbread, said to be native to Vulcan, the home planet of Mr. Spock in the Star Trek show. Star Trek places Vulcan in orbit about Epsilon Eridani, but there is disagreement about this among Trekkies, because Epsilon is such a young star and Vulcans are supposed to be an advanced ancient civilization. Epsilon Eridani does have at least one planet, Jupiter sized.

Fomalhaut Fritters
Mixed corn and potato fried pancakes, with a soy-based sauce.
***Inspired by** Fomalhaut, a bright star in the southern hemisphere. It's distance is 25 light-years. Fomalhaut is visible in the northern hemisphere only during the autumn, and very low on the southern horizon, in a region with few bright stars. It is about twice as massive as the Sun. Fomalhaut is a young star, surrounded by a huge debris disk. A planet appears to be embedded in the debris disk.*

Groombridge Binary Lunch
Tabouli and Hummus
***Inspired by** Groombridge 34, a binary star system, 11.7 light-years distant in the constellation Andromeda. It consists of two red dwarf stars in a nearly circular orbit. Both stars exhibit variability due to random flares.*

Red Star Beans, Beets and Onions
Red beans, red beets, red onions, red peppers
***Inspired by** the star Gliese 581, which is 20.5 light-years distant, in the constellation Libra. This is a red dwarf star, which may have as many as 6 planets. One of those planets could be Earth-like, because that planet is located within the habitable zone, the range of distances from the star where water can exist as a liquid.*

Starman and Friends

I put in an order for the Epsilon Eridani salad. It was OK, but not out of this world. Later on, I sure ate a lot of that salad. It kept me healthy.

I recognized someone and joined him at his table. My friend was a local character who went by the nickname "Starman." He was a white man in his 40's. He had a short black beard, shaggy black hair, light blue eyes and a prominent set of very white teeth. He wore large round glasses with thick black rims.

Starman was full of enthusiasms, but sometimes he seemed not connected to reality. He was very smart, well-informed and capable of coming up with some surprising and inventive ideas. Starman's day job was supply clerk and expediter for the UC Space Science Lab, on the hill above campus. The expediter job took him downtown a lot, and I think he took advantage of that to spend plenty of time at the SETI Café.

With his earphones on, Starman was connected to his laptop computer, staring into space, listening to ET Music.

There were several other people in the Café; most, like Starman, plugged into their laptops. They were also eating one of the star-inspired food items. The Café did a good business.

Starman came back to Earth when he saw me at his table.

"What are you listening to?" I asked.

by Steve Geller

"This is Epsilon Eridani," he replied slowly. "I'm channeling an ET."

"Do you hear a voice?" I inquired.

"No, I just hear the music. But the music makes images in my mind. I'm receiving the thoughts of someone who lives on a planet orbiting around the star. Maybe I'm getting thoughts from several people."

"Listen." He handed me his earphones.

I heard rising and falling hums, some pings, musical tones and plunks. Some of it could be music, but of a very abstract atonal kind. For me, the effect was like listening to an orchestra tuning up. Mostly it was just random noise. I might have heard better music by listening to the honking horns, roaring engines and screeching tires of Berkeley street traffic.

"Do you get ET thoughts when the music is coming from some other star?" I asked.

"Yes, I do," replied Starman. "But not in all cases. I've channeled an ET from Tau Ceti, Gliese 667 and Kepler-22. These stars all send me ET thoughts."

"Are the ET thoughts interesting?"

"Well, sort of, " replied Starman. "What I get is kind of vague and fuzzy." He smiled apologetically, spreading his hands. "These are extraterrestrials, you know; they don't perceive the world the same way we do."

Starman continued, "Some people here think that the human brain somehow extracts information from the ET Music and assembles it into, if not a picture, sort of an impression, a pseudo-memory, a structure of ideas. I suppose the music might stimulate creativity or jolt one's imagination."

by Steve Geller

A SETI Scientist

Starman introduced me to some other Café patrons, telling them that I was a journalist writing about progress in detecting ET civilizations.

I threw out a question to the group, "Do any of you work on SETI projects? This ET music is interesting, but it's not easy to believe that it is coming from an ET civilization. Has anyone actually listened to a signal from an ET?"

A mature red haired woman, who had been regarding me suspiciously, eventually spoke up. "I have worked at the SETI Institute in Mountain View. My name is Dr. Ruth Stern. I'm now with the astronomy department here at UC; I sometimes work with "SETI At Home," the Berkeley group that helps people use their home computers to process radio telescope data, looking for SETI signals.

"As far as I know," said Dr. Stern, "Nobody has yet picked up an ET signal for sure, although there have been some intriguing things received.

Some years back, there was a bit of a flap. Have you heard about the *Wow!* signal?"

I said that sounded familiar and asked her to tell me the details. She began:

> In 1977, a radio astronomer, Dr. Jerry Ehman, was scanning the sky looking for signals. The usual thing is surges of signal intensity spread across a band of frequencies. It is unusual to see a narrow-band signal, a peak at one frequency. If an ET is transmitting, scientists think it will be narrow-band, because that's what Earth's radio and TV signals look like.
>
> In those days, the radio telescope signal was displayed on a computer printer as numbers, 1,2,3...9. The stronger intensities showed as letters A,B,C,D,....
>
> One day, one of the sweeps showed the sequence of numbers and letters "6EQUJ5" -- the 'U' at the peak indicated a signal intensity 30 times background.
>
> This peak was at one frequency. There was no such surge at any of the other monitored frequencies. Using a red pen, the astronomer wrote the comment "*Wow!*" on the printout right next to the surge sequence.
>
> This comment became famous as the name of the signal.

"Wow!" I blurted, rather foolishly. "Was this an ET radio station? Does the *Wow!* signal keep coming in? Do they know what star it was coming from?"

Dr. Stern held up her hand and smiled. "Whoa. It was just a peak in signal intensity, not a message, nothing audio or digital as far as anyone knows. Also, the signal did not repeat. The *Wow!* Signal never appeared again."

by Steve Geller

Dr. Stern continued:

> I should point out that the frequency of the *Wow!* signal was very close to 1420 MHz, the frequency at which abundant cool neutral interstellar Hydrogen radiates. Also known as the 21-centimeter line (the equivalent wavelength for the frequency), this very common background radio frequency had been proposed as the most likely one for ET to use to send a message to Earth, because, since interstellar Hydrogen radiates 1420 MHz fairly steadily, anything above background intensity at that frequency would be seen as unusual, likely to have been generated by an intelligent technical civilization. It would get our attention.
>
> In the early decades of SETI, 1420 MHz was the only frequency at which most observers chose to listen. Today, with much better electronic equipment, SETI listens at millions of different frequencies.
>
> Because the interstellar hydrogen swirls around with the rotation and internal motions of our galaxy, the signal is spread over a band of wavelengths centered on 1420, by the Doppler effect.
>
> One thing that could have made that *Wow!* signal is a satellite of some sort at just the right distance, going just the right speed, to mimic a celestial object traversing the sky. So that's a possibility, but it seems pretty unlikely — it should have been seen by more observers.
>
> Of course, the signal could also have been spurious, the result of an equipment malfunction.

For a long time, Dr. Jerry Ehman, the radio astronomer who saw the *Wow!* signal, accepted the possibility that the signal came from a piece of space debris reflecting a signal back down into the antenna. But he no longer believes that to be the case. Various other sites have tried to pick up the *Wow!* signal, including the Very Large Array in New Mexico -- but no luck.

If the signal did come from ET, then it was not beamed deliberately at Earth; rather, Earth just happened to be in the way as it swung by, like the beam from a lighthouse.

We do know where in the sky the *Wow!* came from – the region of Sagittarius, dense with stars from our galaxy — but we have no idea how far away the source was. If the source was on a planet many light-years distant, then the indicated intensity might have been too powerful to expect any ET to generate, especially if the beam is thought to have been on all the time as the telescope swept by. Of course, that ET might have simply taken a quick shot in the dark, and sent a brief strong transmission in the direction of a likely star system. Perhaps it was a highly compressed burst?

We're cautiously sure that *Wow!* could not be something natural, like a quasar. Natural radio sources spread energy across many frequencies, rather than concentrating only at a single frequency. *Wow!* was definitely narrow-band.

by Steve Geller

It has been noted that SETI research seems to assume that there are many ETs out there eagerly sending us their messages, while we on Earth don't send out much of anything deliberately. The vast bulk of radio emissions an ET might hear from Earth is leakage from our radio broadcasts, radars and satellite communications. If Earth's behavior is any guide to the behavior of ETs, then SETI should expect that the most likely thing we would pick up would be an ET's local radio leakage, not a signal specifically aimed at us.

Remember, ET probably doesn't know we're here.

I and the rest of the audience fell silent for a moment. It was a fascinating story, about a real signal from a possible ET.

Kepler Instrument Finds a Plenitude of Planets

I had another question for Dr. Stern. "So there are a lot of stars with planets out there, right? Isn't it reasonable that a few of those planets would be like our Earth?"

She replied, "There is much we don't know about planets and life. It may turn out that there are some very good reasons why there's only one Planet Earth in our galaxy. We might even be unique in the universe. We just don't know right now. We haven't detected an ET signal, but astronomers certainly have found plenty of planets, especially since the Kepler instrument went up."

She paused, smiled and said, "Yes, I agree with you that there are plenty of places for ET to roost out there. To be suitable for life, a planet has to be orbiting not so close to its star that the planet is too hot, nor so far from its star that it's too cold. For each star, there's a distance range that's just right -- the "Goldilocks Zone." Astronomers and biologists call it the habitable zone -- the region around a star within which planetary-mass objects with enough of an atmosphere can support liquid water at their surfaces. Not every star has a wide habitable zone, and the zone for small cool stars can be so close that the planet can be zapped when the star puts out a flare.

by Steve Geller

The Drake Equation

Dr. Stern then noted: "Planet possibilities can be quantified. Perhaps you know about the **Drake Equation**?"

I definitely had read about it. Dr. Stern began to go over the details:

> The Drake Equation was developed in 1961 by the radio astronomer Frank Drake to use in lectures he was giving on SETI, so that he could give a rough estimate of how many extraterrestrial civilizations we should expect to find in our galaxy.
>
> The Drake Equation consists of many factors, mostly estimated counts and fractions, multiplied together to give a number for technological civilizations that may exist in our galaxy.
>
> One factor is the fraction of stars that have planets. We now know that number to be quite large.
>
> Another factor is the number of planets, per solar system, with an environment suitable for life. We still have only guesses for that one.

Depending on the particular values selected for the factors, the Drake Equation yields a number going from 1 (our civilization is unique) to hundreds or thousands of ET civilizations. Reasonable values suggest that there are plenty of potential places for ET to develop; it is definitely worth the effort for SETI to keep looking. NASA's Kepler Mission was designed to find planets around other stars. At first, it found plenty of Jupiter-sized planets, but now Kepler is turning up more and more planets which could be Earth-like.

The Kepler research team found that 50 percent of all stars have a planet of Earth-size or larger in a close orbit. By adding larger planets detected in wider orbits up to the orbital distance of the Earth, this number increases to 70 percent.

Extrapolating from Kepler's observations and results from other detection techniques, scientists have determined that nearly all Sun-like stars have planets.

And there are the "super-Earths" being found, rocky planets up to twice the size of Earth. Are they too big? We don't know.

by Steve Geller

The Fermi Paradox

Many times, I had looked up at the stars in the night sky and seen the Milky Way. I had studied pictures of big star clusters and of the huge galaxy in Andromeda. Given all those stars, given that 50% of them have planets, to me it seemed unavoidable that many ETs were out there, living on worlds like our Earth.

I wanted Dr. Stern to tell us more, so I asked. "Are we going to find out that all those Earth-like planets are similar to Mars or Venus, with no life? Surely there must be a few twins of Earth, with a civilization capable of space travel – capable of visiting us."

It turns out that I am far from the first to ask such questions. At least one very great mind has examined these issues. Dr. Stern began another story:

> The famous physicist Enrico Fermi, during a lunch at Los Alamos, once asked the simple question "So? Where is everybody?" He thought Earth should have had ET visitors by now.
>
> Surely, Fermi thought, at least one of the extraterrestrial societies would have achieved or passed our technical level of space travel and made an exploratory voyage to us from another star. Fermi pointed out that, even moving at less than light speed, the time required for space ships to cross and colonize our galaxy is shorter than the age of our galaxy.

Since it is a fact that we have yet to receive an ET visitor or an ET signal, the "where are they?" question has become known as the **Fermi Paradox**.

The simplest resolution of the Fermi Paradox is that we clever beings on Earth are unique, or at least extremely rare.

A less-simple resolution is that while there are many technical civilizations out among the stars, we here on Earth happen to be the first to develop our level of technology. We don't hear from anybody else because nobody else has yet figured out radio, let alone space travel.

Our own spacefaring technology has gotten as far as robotic investigation of the outer planets of our solar system, but has yet to produce a launch toward even so close a destination as planets of our nearest star Alpha Centauri. Perhaps no ET spacefarer has progressed beyond our level, or if they have, those ETs are located too far away to get to us. Their transmissions might be coming from so far away that we simply can't detect them.

There would be a big breakthrough in SETI if some technical development made it possible to suddenly pick up signals we had been missing. Clever beings like us could turn out to be not rare at all.

Dr. Stern paused, smiled at her small audience and said:

"So? Then where are they?"

She began to eat her lunch – I think it was the Sirius Sustenance.

by Steve Geller

Felix

Using Starman's computer, I tried listening to the ET Music to see if I could channel an ET. I was not having any success, when Starman beckoned a young man over to our table. He was introduced to me as Felix Fanchot, an Astronomy graduate student. Felix was rather short and stocky, in his late 20s, with curly dark blonde hair and a round pink face. He had a bright smile.

Felix and I began an interesting conversation; he was full of technical information. "What kind of planet is at Epsilon Eridani? " I asked him. "Is it anything like Earth?"

Felix replied, "That star is similar to the Sun, but so far no Earth-like planets have been detected in orbit. There is one big planet, probably a gas giant like Jupiter, but even bigger. There's a lot of orbiting junk there. Epsilon Eridani actually has two asteroid belts. The orbit of that big planet is eccentric, so its gravity keeps stuff stirred up. This makes it unlikely that any twin of Earth was able to form, but who knows?"

"Where is Epsilon Eridani in the sky?" I asked. "Could I find it without a telescope?" "Sure, you can see it," Felix replied. "It's part of the constellation Eridanus, which represents a wandering river. Eridanus begins below and to the right of Orion. Epsilon Eridani is not far from Orion's bright blue star Rigel. Epsilon is visible to the naked eye, but it is a relatively inconspicuous star." Felix showed me this data page from a file on his laptop computer:

by Steve Geller

Epsilon Eridani

Distance 10.4 light-years.

Main Sequence star; Class K2; 0.82 solar mass. Surface temp 5000K; orange color.

Age probably less than a billion years. Has a higher level of magnetic activity than the present-day Sun; stellar wind 30 times as strong. Its rotation period is 11.2 days at its equator. Spectrum shows a comparatively low level of metals.

At least one Planet -- Mass 1.55 x Jupiter , Radius unknown. Orbital Period 2500 days (7 years); eccentric orbit; Semimajor axis 3.39 AU; semiminor 0.72 AU

System includes two belts of rocky asteroids, one at 3 AU and a second at 20 AU. Undetected 2nd and 3rd planet may be maintaining the second belt.

A third, icy ring of material extends from about 35 to 100 AU. (Solar System's Kuiper belt goes from 30 to 50 AU) Epsilon Eridani's outer icy ring holds about 100 times more material than our Kuiper belt.

Having reading Felix's Epsilon Eridani file, I had a few questions.

"I know that a light-year is a distance; it's how far light travels in one year, right? So what's an AU?"

Felix explained:

> AU is the abbreviation for Astronomical Unit. It's the average distance between the Earth and the Sun, about 93 million miles. We use the AU to talk about distances on the scale of the solar system. For example, the distance from the Sun to Mars is 1.5 AU; Sun to Jupiter is 5 AU. Neptune is 30 AU. Far-out Pluto has an eccentric orbit, so its distance to the Sun varies from 30 to 50 AU.
>
> We use the light-year to describe the much larger distances between the stars. It takes 63,241 AU to make a light-year. The stars are very far away; the nearest star, Proxima Centauri, is over 4 light-years distant.

by Steve Geller

I had another question. "What does it mean to have a 'low level of metals'?"

Felix answered:

"Astronomers use 'metal' to mean any element heavier than Hydrogen or Helium. I think the terminology comes from the practice of using the star's spectrum to measure the abundance of Iron, and using that metal as an estimate of the abundance of all other heavy elements. In order for a rocky, iron-cored planet like Earth to form, the leftover debris from forming the star should have plenty of metals. By the way, all 'metal' elements got formed long ago by nuclear fusion inside large stars, and they were spread over a large region of space when those stars exploded as supernovas."

"OK, Felix," I said. "Thanks a lot. I have just one more question. You say the surface temperature of this star is 5000 K. I know about F for Fahrenheit and C for Celsius; in what degrees system is K?" Felix quickly began another explanation:

> K means Kelvin.
>
> The Kelvin scale was established by the British physicist Lord Kelvin and is used for absolute temperature. K has the same degree-steps as C, but zero degrees K is absolute zero – the coldest possible; heat vibration has ceased. The freezing point of water, zero degrees C, is 273 degrees K. Room temperature, about 20 degrees C, is 293 degrees K.

The Sun has a surface temperature of about 5700K. The surface temperature of Epsilon Eridani is about 5000 K, which is 4727 C.

Want that in Fahrenheit? Work it out.

Felix wrote the Celsius to Fahrenheit formula on a napkin:
F = (9/5)C + 32

I pulled out my calculator and soon announced "8,541 degrees F. Very hot indeed. OK, now I think I understand the technical basics."

"Something else to keep in mind," Felix added:

> The gravity of large planets in a system vacuums up much of the miscellaneous comets, asteroids and rocks, which would otherwise bombard a small rocky planet.
>
> There were several mass extinctions of life on Earth caused by such bombardments. The most famous is the impact 65 million years ago which might have wiped out the dinosaurs.
>
> The gravity of Jupiter, and Saturn too, is thought to have captured enough large objects early on so that only a few of these extinctions happened to Earth.
>
> The same thing might take place in the planetary system of another star, removing bombarding rocks from an orbit region, allowing a small rocky planet in that region to develop life going without an impact wiping it out.

For this reason, star systems which have no gas giant planets probably have little chance of developing a life-bearing planet like Earth.

Listening to ET Music

I turned from Felix back to Starman. "What do you hear in the ET Music?" I asked.

"From Epsilon Eridani? Nothing specific, really; I sure don't hear anything like 'tell your leader we are here and ready to do business.'" He gave me a toothy grin.

I asked Felix. "Starman says the music makes images in his head. Does this happen to you?"

Felix shrugged. "I suppose any music makes images in my head. I get vague impressions, which could possibly be the thoughts of extraterrestrial beings, but if so, I don't understand much from them. You are aware that this ET music is artificial, aren't you?"

That stopped me for a moment. I asked, "You mean it's not really coming from the stars?"

Felix shook his head impatiently. "No, I mean that the radio signal from the star is being artificially processed to produce the music you're hearing."

I probed Felix for more information. "I understand that the signal from the star is received by a radio telescope, as high-frequency. The high frequency is reduced to bring it into audio range so we can hear something. Is this what you mean by artificial?"

"Yes, but there's more involved than a shift in frequency," replied Felix. "The signal is not a pure tone; there are lower frequencies riding on top of the high-frequency signal. This is called modulation. The signal processing converts the lower frequencies to musical tones. The lab that does this processing tries various settings until they get something interesting or pleasant to listen to, for that star. Then they send the audio result to the website."

I asked, "Is it always the same processing?"

"No, it's different for each star, and the processing used for each star is changed from time-to-time."

I pursued. "Do these signals mean we are in contact with an extraterrestrial intelligence?"

Felix chuckled and gave a shrug. "Not necessarily. There are plenty of interesting signals generated naturally by any star and the stuff orbiting around it. If we processed pressure waves from Earth's weather, we could probably generate some kind of music. The surface of a star can pulsate; that could generate an audio signal. "Frankly, most radio astronomers scoff at ET Music as irrelevant to science, but many radio telescope observatories do supply signal samples for us to process, so that the SETI Café can provide a wide variety of ET Music.

"But, you know," Felix brightened, "There could be ET radio and TV mixing with the star signals. The trouble is that, because of the many light-years of distance, such signals would probably be too faint and full of interference to make sense of.

"I think it's simply fun to listen to the ET Music that we generate, without trying to make mysterious messages out of it."

I talked to several other people. One man told me that he was in business, and would like to begin commercial trading with an ET civilization. I explained how exchange of hard goods is complicated by the many years delay due to the distance of the stars. But I encouraged him to look into marketing whatever soft goods, information and culture, we might receive in a message from an ET.

by Steve Geller

Music from the Void

I asked Felix, "Well, anyway you always know where the ET music star signals are coming from, right?"

"We know precisely where in the sky the radio telescope was pointing, and the time, but there might be more than one object in the view, at various distances from us."

Felix called over to someone who had been talking with Starman at another table. "Hey, Cosmo, has anyone figured out what star is sending that 'Music from the Void'?"

His question was directed to a tall slim young man with short dark hair. As we went over to their table, Starman told me: "This is Cosmo, the Café manager."

Cosmo told us, "Last I heard, it's still from a hole in space. The astronomers say there's no nearby star at the location, and the music doesn't always come through coherently. It might be the signal from an Earth-bound radio station reflected from some orbiting junk."

"What are you guys talking about?" I inquired.

Cosmo explained, "We call it Music from the Void -- or MV for short. The Stanford radio telescope people came across it when they were doing some calibration tests. They pointed their telescope at what they thought would be a blank quiet spot in the sky, coordinates where optical maps show no bright stars and radio maps show just the steady hiss of Hydrogen and the cosmic microwave background.

"They ran the test using a fairly large number of frequencies and didn't pick up much of anything. But then at one frequency, they got a steady signal. They re-visited the coordinates and frequency several times, and nearly always found the signal present. One of their technical people tried an AM audio demodulation, and out came this mysterious music."

Cosmo asked, "Starman, have you got any MV files on your computer? Would you play one of them for us?"

Starman fiddled with his laptop, then handed me his earphones. I began listening to the Music from the Void.

It started with a steady hum tone, like a bagpipe drone, but with no tune on top of it. Then the tone slowly varied in pitch, up and down. I was about to hand back the earphones when a second tone entered, at about the same volume level, its pitch changing independently. Then a high pitched tone began to come through; it lasted more than 10 seconds or so before going away. Several other tones came and went.

The overall effect was a rather soothing simple kind of music. There was no tune that I could recognize. Some of the tones had a timbre like a flute or an organ pipe; others had overtones similar to those of a violin or an oboe.

The tone mixture, each component rising and falling in pitch independently, combined to make a harmonic melody. It was very pleasant.

Cosmo was grinning broadly, much amused by my awed and mystified expression as I listened on the earphones.

Starman then told me, "Several very musical people have listened to MV, but nobody has recognized any of the music. They might recognize some of the instruments: one could be a flute; another might be some kind of stringed instrument. It sure isn't J. S. Bach or the San Francisco Symphony. It's not always the same music or instruments, but some do repeat, and as I said, we don't get the signal all the time or with this kind of quality. So we record the good MV stuff when we can."

I could hear gaps in the music, and occasional static and warble distortion. "Could this be an artifact of ET music signal processing?" I asked.

Cosmo said: "Actually, MV does not go through the processing we use to produce the ET music. All that has been done is to demodulate the signal and bring out the audio. Maybe there's been some additional filtering; I don't know."

Cosmo continued. "We asked around at several local radio stations, giving them the times when we had heard the music, but none of them were broadcasting any hum tone melodies like we heard at those times. Anyhow, the carrier frequency isn't particularly close to that of a local radio station. I suppose it could be some kind of bounce from a station far away. We at the Café have decided, for now, not to advertise MV as one of our ET music offerings. Only a few Café customers know about it.

"There's one guy who is trying to clean up some of the recorded MV, to get rid of the gaps and distortion. He wants to sell CDs of Music from the Void. Neat idea, I think."

Starman took over from Cosmo to tell me about another aspect. "MV can affect some people very strongly; they kind of go into a trance and have a mystical experience. I think this comes from listening to MV for too long at a time. I've listened to quite a lot of it, but so-far, I don't get any mystic visions." He gave his toothy grin.

by Steve Geller

Other Prospects for Contacting an ET

Over the next few days, Starman introduced me to several other patrons of the Café, mentioning my SETI journalism project.

I should apologize at this point for not always reporting conversations or conference talks in detail, but rather just giving the substance of what was said. Many times, I couldn't take notes fast enough; I just got overwhelmed. It shouldn't be a problem. I don't think I missed recording anything important. I sometimes used a pocket voice recorder. It worked fairly well for presentations at meetings, but it picked up too much distracting noise in environments like the SETI Café.

Anyhow, all the other patrons told me that listening to ET Music gave them mental images and vague ideas about far-off places. When I suggested that most music has that kind of effect, I was told that the images and thoughts from ET music are unique. These customers at the SETI Café thought they were really listening to murmurs from a far-off alien civilization, some of the time, anyway.

One patron surprised me by likening the ET Music experience to listening to short-wave radio. "There is structure in it," he said. "I think I hear fragments of Morse code, of music and voices fading in and out."

"You hear words?" I inquired.

"I don't myself hear any recognizable speech," he replied. "But some of my friends think they hear some voices."

Starsong

"By the way," he added, "There's a musical group around here that plays stuff they say is derived from our ET Music. Their name is "Starsong." The group has a girl singer, who composes words to go with the music. They're kind of a chant. Some are in English; some are in other languages, like Hebrew and Sanskrit. I don't think she claims to actually hear the words of these chants in the ET Music."

Starman spoke up: "I've attended a concert given by Starsong. They're good. They very cleverly made the hums and plunks more musical, and also added some rhythm. I really liked it."

ET and Religion

Starman called to another Café patron. "Hey, Father Paul, tell us about ET and God."

As the man came forward, Starman announced to everyone "This guy thinks that God sends revelations to us from the stars."

A solemn-looking older man came over to our table and introduced himself. "I'm Father Paul Beni, a Berkeley Catholic priest." He was grey fringed, balding; he wore square black-rimmed glasses. He had on a black shirt with a roman collar. He had a warm and peaceful smile.

Father Paul began, "To me, it seems very likely that messages from the stars might include messages from God. If we believe God is a universal spirit, then God is everywhere in the universe, and communicates with ET civilizations at distant planets just as with us.

"When we finally find life on another planet, it will have gone through a very different origin process than ours, and will turn out to be quite different from life on Earth. We might not even recognize it as life."

A person at the next table spoke up: "Perhaps Jesus visited other worlds. The Mormons say that Jesus visited the North American Indians."

A student at another table identified himself as Muslim. He said he thinks that Islam is OK with ETs. He told us: "There are several verses in the Koran in which God's creatures of various kinds are mentioned. There is one verse which goes like this:

> *Among His signs is the creation of the heavens and the earth, and the living creatures that He has scattered through them: and He has power to gather them together when He wills.*

"Based on that verse, and others, some scholars of the Koran say Muslims should be free to believe in the existence of life on other planets."

The student added, "Note the reference to living creatures scattered through the heavens. That could be about creatures on planets other than Earth."

Another man broke in, speaking loudly. "There's another possibility. The ETs might have an entirely different God."

Father Paul agreed. "God worshipped by ET indeed might look very different from the God we know. Historically, God has shown Himself in many different aspects."

by Steve Geller

Somebody else interjected at this point: "It could be that God is an ET!" Everyone turned to look at the speaker, who grinned. "Don't the Mormons also say that the throne of God is located on some far-off planet?"

Yet another person broke in, expressing impatience. "I can't buy any of this! If we really do share God with a bunch of ETs, then why hasn't God brought us into communication with the ETs? Aren't we all God's children?"

Dr. Stern gently applied the brakes to this cosmic theological discussion by saying, "Well, we'll find out if and when we get to carry on a conversation with an ET. Keep in mind that the nearest possible ET is ten or more light-years distant. This means message exchanges will take decades. It won't be much of a lively discussion, with many years passing between a message going out and a reply coming back.

Dr. Stern added, "Also, as Father Paul points out, there's no reason to expect ET life forms to closely resemble humans. And even if ET is intelligent, resembles us and sends messages, the language system ET uses might be so totally different from ours that communication isn't possible. Think of how different some Earth languages are. Look at English, Chinese and Navajo. Decoding and translating an ET message may turn out to be very difficult.

"And there's yet one more possibility, something somebody suggested to me just last week. The reason we don't hear their radio messages is that all the ETs have gone digital, or they use cable systems instead of broadcasting." She grinned.

Someone in the back raised a hand for attention. "Yes, Karen?" Dr. Stern invited a young woman to come forward and be introduced. "This is Karen Banks. She's a graduate student in linguistics. She attends some of our SETI seminars and has contributed some useful insights. Karen, what have you got to tell us?"

Karen was a diminutive young lady with short brown hair. She began shyly, "Since we have yet to make any contact with an ET, we should not assume that any ET will think the same way we do. Some humans are very good at learning and retaining languages, while others can barely manage the language they were born to."

She paused, looking around at the others uncertainly. "It may be that one or more of the ET civilizations is very good at linguistics. They may have been able to figure out our English. We'll know that when an ET starts sending messages to us in English."

by Steve Geller

That made me think: I suppose it is faintly possible that some really clever ET will not only figure out our language, but show us how to use language in such a way that anyone can understand us. If we can figure out how to send messages in an ET-comprehensible way, we could translate our Wikipedia into it, package it and beam it to where the ETs are. Right now, Wikipedia articles are available in over 40 different languages, including Shqip (Albanian), Cymraeg (Welsh), Hrvatski (Croatian) and Esperanto.

Would we recognize an ET writing system as transmitted in a message? How would their alphabet be encoded?

Felix, who had kept silent during all this talk, tried to bring us back to reality and practicality. He said:

> Because the universe is very large, with billions of galaxies made up of billions of stars, and because recent observations have shown that plenty of stars have planets, there's nothing that compels us to accept that life is vanishingly rare. ET might turn out to be very common, inhabiting a wide variety of environments, using many different technologies and ways of communicating.
>
> Perhaps extra-terrestrial life won't be an exact duplicate of what we see on Earth. Here is a very simple and clear definition of life that doesn't call for life to be rare: life is a self-regenerating, self-adapting system.
>
> The development of life could well be one of the common processes of the universe, like star and planet formation.
>
> Right now, it looks like our living planet is unique; we won't know any different until we encounter ET life — an example of life other than what we have on Earth. How different will it be?

by Steve Geller

There's a distinct possibility that Earth has been an extremely lucky member of the universe's sizeable collection of rocky wet planets. Geologic evidence shows that our planet has suffered numerous environmental disasters (snowball Earth, hot earth, asteroid impacts) and life on Earth has undergone some major extinctions. Earth life has avoided permanent extinction by surviving a large number of close calls. Other rocky planets could have developed life, and then lost it to a disaster. Earth may be the last planet standing.

We really don't know whether life is inevitable, on any planet similar to Earth, or whether life is the result of an extremely lucky, highly improbable sequence of events and conditions, which may have happened only once. Some people call this the "Gaia vs Goldilocks" dilemma: is our Earth situation an inevitable outcome for certain kinds of worlds, or a unique outcome on the only world that was "just right?"

I'd like to go for Gaia.

Life probably once existed on Mars. There is evidence from rock formations photographed by the Mars rovers that in past ages, our neighbor planet had lakes, rivers and maybe even oceans. But there is no life there today — well, it's still possible that hardy native Martian microbes remain from the water epoch. Unfortunately, it's all too likely that the Mars landers arriving from Earth have deposited some of Earth's microbes, so it might be hard now to tell now what's native to Mars from what's not. We're fairly sure that there are no intelligent beings on Mars, who can send us radio signals. They sure haven't done so, anyway. None of them have met our landers. Not even Martian insects have come buzzing or crawling around our landers, a reception that certainly would have happened if an ET spacecraft had landed on Earth.

Marine life might exist right now on Jupiter's moon Europa or Saturn's moon Enceladus; both bodies show signs of an ocean of liquid water beneath an outer crust of ice, the water being kept warm by the big planet's gravity, squeezing the body of the moon. There might even be some form of life in the swirling clouds which constitute the atmospheres of Jupiter, Saturn, Uranus and Neptune, but so-far, such life has not been detected.

The main contention of the Gaia hypothesis is that life, however it originates, creates and sustains conditions suitable for life to continue. Life itself may very well be a big contributor to stability of climate on an Earth-like planet.

Regardless of all this, the question driving SETI isn't just about life. It's whether thinking beings like us humans are out there, with anything like our technical and language capabilities. Life of some kind is almost certainly out there, but human life may still be unique, a rare fluke of nature. There simply may not be anyone out there we can talk to, relate to and learn from.

by Steve Geller

But we still keep listening with SETI.

I had been so intently listening to this fascinating talk and furiously scribbling notes to record all the great ideas, that I hadn't noticed how late it was getting.

It came as a mild shock to us SETI enthusiasts that the SETI Café might want to close for the day. The Café staff was beginning to clean up the tables and stack the chairs. The Café manager was very nice, but firmly herded us toward the door.

I remembered thinking, as I departed, that most of the people I had met were very interesting and well-informed. Some of them really deeply understand the issues involved in SETI.

I had planned to do some on-campus interviews, but I thought that for now, I could do just as well talking with people who came to the SETI Café.

A Philosophical Doubter

Having departed the closed Café, I began to walk back to where I was staying, intending to remain awake for a while and write up my notes.

For a change, Berkeley's Bay Area night sky was clear of fog. I could see familiar northern hemisphere constellations: Ursa Major (the bear/dipper/plough), Lyra (the harp), Cygnus (the Swan), the great square of Pegasus, the W-shape of Cassiopeia.

I imagined an ET on an extrasolar planet, doing just as I was, looking up at his own night sky pondering the same things that I was. I've been told that in the sky of the planet thought to orbit the nearby star Alpha Centauri B, the Sun would appear as a bright yellowish star in the constellation Cassiopeia.

While I was enjoying my star gazing, a man came up behind me and touched my arm to get my attention. I remembered him as one of the group at the SETI Café. He was in his 20's, tall with a thin pale face and longish brown hair. He wore rimless glasses.

by Steve Geller

"Hi," he said, "I'm Dave Starkey. I'm a grad student in Philosophy. I hang out with the SETI folks sometimes and try to keep them conceptually honest. Tonight they got way out of hand, especially those religious people. Speculation is fun, but I think they're letting the fun get too far ahead of the facts. We have yet to detect an ET, let alone study its language. Whales, birds and bees are at least distantly related to us because we all developed as part of Earth life, but any extraterrestrial will not be related to us at all. If we can't talk to any of our fellow Earth animals, we are not likely to have a successful discussion with an ET."

"I see your point," I said, "but the building blocks of life seem to be widespread in the universe. ET may turn out to be surprisingly like us after all. We may find out someday, but for now, you're right; we don't know."

I gave philosophical Dave a pleasant smile and told him, "I see no reason to totally write off the possibility of ET communication."

Dave frowned, and then put on a sneer. He began to lecture me:

> I have a simple resolution to the Fermi Paradox. ET is already here! Or maybe ET has paid us a visit in the past, and we are incapable of recognizing the evidence, past or present, because we don't have the correct concepts. ET might be visible in plain sight. ET might even dwell within us, as an alien-introduced alternate personality.
>
> Even if ET has never been here physically, I'm quite sure that SETI has been receiving ET messages for a long time, but because the ET's have developed such totally different communications technology, we don't recognize what we're receiving.
>
> One rather extreme possibility is that the ETs out there don't even use electromagnetic waves, and we'll have to wait until we make the same physics discovery they did before we can hear their transmissions.
>
> SETI may be a big waste of time. I think SETI researchers should try more methods of communication, not keep hopefully listening on the same old frequencies.
>
> My guess is that Earth-like technical civilizations are sufficiently rare and thinly distributed that we are unlikely to encounter ET, either as a physical visitor or sending us a radio transmission. The galaxy is just too big and the technical issues too daunting.

by Steve Geller

All we can hope for is a spectacular discovery which either allows us to detect ET radio or execute rapid interstellar jumps. I'd even be pleased with the development of some means of transporting small quantities of matter (or just patterns) fast enough to act as an interstellar telephone.

I asked Dave what evidence he sees for visits from ETs, but he just kept ranting about our inability to understand, and that some of us humans may be aliens.

I finally broke loose of philosopher Dave and continued on to where I was staying. Some of Dave's ideas might make sense. He's right that for now, our speculation is way out in front of the facts.

Cosmo

The next day, I did an interview with the manager of the SETI Café, a man I'd met earlier during a SETI discussion at the Café.

Cosmo Gomez was a tall, slim, handsome Earthling, with short dark hair and a light olive complexion. He somewhat resembled the character Mr. Spock from Star Trek, but Cosmo was much more emotionally alive and did not have the funny Vulcan ears.

He was wearing a T-shirt which showed a spiral galaxy, presumably the Milky Way. An arrow pointed to a location on one of the outer bands of stars. The legend read: "You are here, Berkeley, California."

I suggested that "Cosmo" might be a stage name for his job, but he assured me he is really called Cosmo. "My mother was a history professor, specializing in the Italian Renaissance." he said, "She told me I was named after Cosimo de' Medici, the Renaissance banker." "OK," I agreed. "Cosmo it is, serving Berkeley patrons cosmic food and ET Music. How many people work here at the SETI Café, Cosmo?"

"There are four of us. I'm the manager. Abner Katz is the head cook and executive chef." He indicated a short, dark, mostly bald, sullen-looking man working behind the order window in the kitchen.

"Doris Chen is our cashier and bookkeeper." The cash register was on a small table near the Café entrance. Seated behind it was a middle-aged Asian woman, who smiled at us.

by Steve Geller

Cosmo indicated a young man working in the kitchen with Abner. "That's Pablo Pino; he helps Abner with the food, gives him a lunch break at 2 pm, operates the dishwasher and handles the receiving and storing of food and other supplies.

"Pablo, Abner and I do the clean-up at the end of the day. Doris takes care of all of our paperwork.

"This restaurant is self-serve; nobody waits on tables. You give your order to Doris and pay her. She gives you a number and passes your order to Abner, who puts your food on the shelf in the order window, where you can pick it up when you hear your number called. You bus your own table and stack your dirty dishes over there. It's an informal operation. Pablo and I help out some in the kitchen, but Abner doesn't like anyone except him doing the cooking." Out of the corner of my eye, I saw Abner nod his head firmly at this point.

I picked up a menu and waved it at Cosmo. "Who did these great astronomical descriptions?" I asked. "Was it Felix?"

"Felix did a few of them," agreed Cosmo, "But most were written by a UC astronomy professor, Dr. Ruth Stern, who comes in here a lot. She's very interesting. You should talk to her."

I said that I'd met Dr. Stern, and planned to do an interview with her.

I continued the Cosmo interview, asking him, "This broccoli and nut salad: you say it's from Epsilon Eridani? Did the broccoli actually come from out there?"

Cosmo frowned and gave a dismissive wave of a hand. "Of course not; Epsilon Eridani is too far away to have any exchange of goods, especially perishable foodstuffs."

He pointed to the menu. "Please notice that the description of every menu item says 'inspired by.' All of the places that inspire our cuisine are located very far away, light-years distant, out among the stars."

Cosmo grinned. "We definitely do <u>not</u> have an interstellar food supplier."

"Actually, our food raw materials are all quite local; I don't think anything comes from farther than Fresno. We're talking about a style of cuisine, not where the raw food comes from. Look, when you visit the Chinese restaurant down the street, do you ask if all the food makings come from China? Maybe a few spices might be imported from there, but the broccoli, chicken and rice were all produced right here in California."

Cosmo picked up a menu and flourished it. "All ingredients for all the dishes in here were produced on planet Earth. But when you try one of our dishes, we hope you will find the experience out of this world." He grinned confidently.

I asked, "How did you decide this was an Epsiloni-style dish?"

He replied, "We got the idea for the salad from one of our customers, who listened to music from Epsilon Eridani. He claimed to have channeled a salad-making ET who told him the ingredients."

by Steve Geller

Cosmo observed my dubious expression. He grinned and said, "Hey, you can travel to Epsilon Eridani yourself and check it out, but it would be a long trip if you're using anything as slow as NASA's current spacecraft. Or you can put on your earphones and hear directly from Epsilon Eridani yourself, using the ET Music — if you have the same talent as our customer. He, and some others too, think they have channeled the thoughts, likes and dislikes of the beings who live, cook and eat on a planet orbiting Epsilon Eridani.

"The SETI Café is just about entertainment and good food, but I won't deny that it is fun to pretend we have a cosmic cuisine connection."

I asked, "Isn't Alpha Centauri closer to Earth than Epsilon Eridani? Why don't you have some Centauri-style dishes on your menu?"

Cosmo shrugged. "We don't seem able to channel anything from Alpha Centauri. It might be that there is no intelligent life there. It is a multiple star system; I'm told that there is at least one planet." He shrugged again.

I persisted: "What about planets that are not Earth-like? Could they have a cuisine?"

He chuckled. "They better be Earth-like if you expect the cuisine to be edible by Earthlings.

"Actually, Abner, our cook here in the restaurant, has a theory that the ET beings are very different from us, but somehow in the channeling process, because it goes through two minds, a kind of translation takes place from what they like to eat in their environment into what we can cook and enjoy here. For example, the broccoli might correspond to some quite different plant that grows at Epsilon Eridani. Enough elements of their cuisine must somehow correspond to elements in ours in order to make the translation work. We probably could not eat most of the actual Epsilon Eridani food; whatever corresponds to broccoli might even be poisonous to us."

That theory made me think of another question. "So it appears that you can only channel successfully from an Earth-like planet, or at least a planet whose pattern of life bears some resemblance to that of Earth? Have you tried tuning in a world like Jupiter? Have you tried any of the other solar system objects where there might be life? What about Jupiter's moon Europa, Saturn's moons Titan and Enceladus? Europa probably has a salty ocean under its ice crust, where marine life may have developed. Same for Saturn's Enceladus; it spurts water geysers; must be quite warm in there."

Cosmo gave me a faintly apologetic smile. "I don't think we have tried any of those places. Are radio telescopes listening to Europa or Enceladus? I know there are radio waves coming out of Jupiter and Saturn.

"No, we don't get any ET channeling from within our solar system, not so-far anyway." He spread his hands. "We can try."

"Another thing," I pursued. "Are there any meat dishes here?"

by Steve Geller

"Good point," Cosmo responded promptly. "Indeed, all our SETI Café dishes are vegetarian. This is where Abner's translation theory might work better. Perhaps no ET animal life forms match anything here, so we don't get any translation. It might also be that we need to try harder with our channeling. Maybe we could channel an ET beefsteak. But meat would make our food more expensive." Cosmo grinned. "Abner says that if anyone turns up edible life on Mars, he's ready to offer such dishes as Martian Meatballs or Martian Microbe Salad. They sound, delicious, don't they?" Cosmo grimaced.

"So," I said, "The whole idea of eating like an ET requires some faith? Doesn't the US Food and Drug Administration come after you?"

Cosmo gave a firm, negative head shake. "Our food is just as safe and nutritious as that in any other inspected Berkeley restaurant.

"You may well need some faith to accept that ET music really can be used to channel ETs. We don't guarantee anything; anyone can give the tune a try."

Cosmo continued, "We are not telling lies, any more than is the car salesman who likens the ordinary car on his display floor to a sports car. We might be selling some fantasies, but we think we are really channeling ET cuisine."

Each to his own fantasy, I suppose. This made me recall, that morning, seeing a big tour bus coming down Bancroft Way. The company name displayed on its side was "Galactic Transporter." Were they advertising interstellar travel service? I think not.

Cosmo went on, "Yes, astronomers do get some radio transmissions from Alpha Centauri, but when we process them into ET Music, we get nothing that anyone, as far as I know, has been able to channel.

"Somebody does the same thing for Epsilon Eridani and we get a broccoli and nut salad. We get fritters from Fomalhaut, and many others." "How about bagels from Betelgeuse?" I suggested facetiously.

That got a grin. "Right, we do have the Betelgeuse Beet Treats." Cosmo pointed to the menu item. "I have to admit that this dish was not actually channeled. We just thought it was a cute name. Same deal for 'Sirius Sustenance.'

"And of course the Vulcan Kreyla Flatbread was inspired by Star Trek. We haven't heard that they have the name copyrighted; if they do and someone objects, we'll stop offering it. Anyway, we don't sell very much of Mr. Spock's home food. I suspect that, for the occasional order we do get from a Trekkie, Abner may just put out some store-bought Scandinavian crisp rye bread. I don't ask."

Cosmo continued, "Yes, I wish we could channel something from Betelgeuse, but that place is not at all Earthlike. Betelgeuse is one of the biggest stars there is. If Betelgeuse replaced the Sun, its outer layers would be past the orbit of Mars; Earth would be barbecued inside Betelgeuse. Really big or bright stars are not likely to have any Earth-like planets.

"I read that Betelgeuse pulsates; it has roiling convection cells beneath its surface. Astronomers can see two pulse frequencies, one with a period of about one Earth year and the other of about six years.

by Steve Geller

"I'm told that pulsations of the surface of a star are fairly common. The surface of the Sun shakes from 'starquakes.' I read where one astronomer described the surface vibrations of various stars as a celestial symphony in which the smallest stars are flutes, the medium-sized ones are trombones and the giants are reverberating tubas. It's very poetic.

"Johannes Kepler, the seventeenth-century astronomer after whom the Kepler spacecraft is named, theorized that Earth and all the other known planets each made their own sound — an arrangement that he called the music of the spheres. Cool, huh?"

I ran my finger down the Café menu. "I see dishes from Tau Ceti, Groombridge 34 and Gliese 581. Those are all relatively nearby stars?"

Cosmo replied, "Yes, they are. But those last two, Groombridge and Gliese (he pronounced it GLEE-zuh), may not be reliable channeling prospects. Dr. Stern tells me that both those stars are red dwarfs. She says they might have habitable zones close-in, but red dwarfs have a tendency to put out flares, sudden bursts of energy, often including X-rays, which would be hard for Earth-like life to deal with."

I asked, "So it's possible that, for example, that the broccoli and nut salad might not be native to Epsilon Eridani at all?"

Cosmo waved his hand, dismissing the idea. "Oh, there's something there. Whatever corresponds to the broccoli and the nuts on Epsilon Eridani might look very different if we could see it here on Earth.

"But we serve the Earth eatable version. We sure do sell a lot of that salad. Epsilon Eridani is much like the Sun. The system has at least one planet. There's plenty of planet-forming material -- it has two asteroid belts. There might be plenty of planets with salad-eating folks out there."

I asked Cosmo to describe how the ET Music data processing is done. He referred me to Dr. Ruth Stern, the UC professor who had composed most of the star descriptions on the Café menu. Dr. Stern teaches in the UCB Astronomy Department and does space science research. I phoned her and arranged an interview; she wanted to meet me at the SETI Café.

by Steve Geller

Dr. Stern

Dr. Stern was waiting for me, the next day, when I arrived at the SETI Café for our interview; I found her sitting at a table, talking with Cosmo, who introduced me.

"Ruth, I think you'll like this guy. He seems to be a nice, honest, unbiased journalist."

Dr. Ruth Stern was a tall, lean woman, about 55 years old, with a somewhat florid face, prominent nose and dark red hair, which was piled high on her head.

We shook hands. Dr. Stern went over to Doris to put in an order for some Tau Tofu and then got herself a cup of coffee. I gave Doris an order for my usual Epsilon Eridani broccoli and nut salad.

"Dr. Stern, what is your connection with the ET Music?" I asked.

She explained, "I help supervise the lab on Campus which receives and processes the signals for ET Music. A couple of my students do the software programming and the tuning of the music extraction algorithms."

I asked, "Do you work with other researchers on this? Do you get any funding from National Science Foundation?"

Cosmo, sitting next to us, gave a soft cough and a small smirk.

Dr. Stern grinned at Cosmo and revealed to me: "Frankly, the science establishment doesn't take any of this ET Music stuff seriously. We do get some donations from interested private sources, but no government funding at all. This stuff is too far out even for SETI. Basically, very few scientists accept that we can channel anything from ET Music, nothing that results in finding out what is going on at the star."

"Few?" I said hopefully. "Do some scientists go along with channeling?" Dr. Stern looked uncomfortable. She said, "I'm not sure I take it seriously myself. I certainly don't write any scientific papers about what goes on here at the SETI Café. We do have a couple guys who are studying the channeling reports, but I think they're writing a paper about urban folklore."

Cosmo broke in somewhat anxiously: "Keep in mind that the SETI Café is providing entertainment. Our customers are free to imagine whatever they wish, as long as they spend money on our food." He smiled brightly. "Any contact with extraterrestrial beings is an added benefit."

"So you don't believe any of it?" I challenged Cosmo. "Hey," he replied, shrugging his shoulders and spreading his hands, "I'm not a scientist. I just manage a restaurant. I try to keep the customers happy so we make a buck. I suppose I do sort of believe some of the ET music channeling stuff. I'd like to. I can keep an open mind."

I asked Dr. Stern. "Is there any prospect of receiving actual ET messages from stars, not just music artificially derived from star noises?"

"I think so," she replied. "We should be able to hear some ET radio, TV, radar and satellite transmissions. It's possible that we could tune in to an ET Internet, depending on how much leakage there was from an ET satellite network.

"Remember that ET communication works both ways. Beings on the planets of other stars may have already developed the ability to detect the abundant radio signals Earth has been spewing for years.

"On the other hand, the reason we don't hear ET could be that our technologically capable life form here on Earth is unique in the universe. But I doubt this. I think the more likely reason is that the many light-years of interstellar space between ET and us are full of things that attenuate, distort and interfere with radio signals. It's really hard to hear clearly from another star.

"This situation could change in the near future. There's a research group here working on some technology to enhance interstellar signals.

They call it **Statistical Signal Enhancement (SSE)**.

"The SSE lab here at UC received a contract from the National Security Administration, to develop ways of recovering broken-up, low-level, irregular or distorted signals. The lab found it convenient to test what they were developing by using it on radio telescope data. This led to developing the SSE for SETI. The SSE is a rather complex system with both hardware and software components.

"When used for SETI, the SSE might be able to take an ET radio signal which has been distorted and broken up during the long haul over the light-years, and put it back together. I know they have been working with some of the same ET signal sets that we use to produce ET Music."

I asked, "How far along is this SSE development?"

"They're getting there," replied Dr. Stern. "They've had good results from various signals they can get by listening to short wave broadcasts. They've produced usable signals from listening to fragmentary fire and police radio frequencies from very far away. I haven't heard that they have produced worthwhile results from any star."

by Steve Geller

Messages Arrive from HD85

Evidently the SSE development was further along than Dr. Stern realized.

A few days later, when I entered the SETI Café around noon, everyone was listening to a new sound from the restaurant speakers.

It sure wasn't voice or music. It was a repeated sequence of a short fast warble, followed by a longer ragged buzz. From my shortwave listening experience, I thought I recognized a digital packet or a FAX transmission.

Cosmo saw me standing listening at the doorway. He came over and gestured up at the speakers. "This is the latest thing from the stars," he told me. "It's a real signal, not something produced artificially. The SETI people are pretty excited. It's coming from some star that they haven't processed before, and it repeats. The radio telescope gets the same kind of signal for about 15 minutes every couple months. This is the latest; we're running it on a loop."

I saw Felix, the astronomy student, having lunch at a table in back. I went over to ask him, "What is this noisy new stuff?"

Felix replied, "You remember Dr. Stern telling you about the technique of Statistical Signal Enhancement? Well, they actually got it to work to some extent. When they tried it on HD85, they got a good signal on one frequency. They tried various ways to decode it, but didn't come up with anything that worked until they got a hint from Bob Weiss."

A Conceptual Breakthrough

One of the great things about the Internet is that any problem can quickly and easily be put before a vast number of minds. Maybe a downside is the foolish fancies that some people come up with, based on little more than wishful thinking. But a major upside is that it is possible for some very clever person to study the data and provide a major insight for understanding.

This had happened with the HD85 messages. A retired engineer named Bob Weiss had obtained some HD85 data files over the Internet from the SSE lab. Working at his home, Bob ran it through some of his own equipment and software. He recognized bits, encoded by a shift between two close radio frequencies: one frequency representing 1, the other frequency representing 0. This digital transmission scheme is called Frequency-Shift-Keying, or **FSK**.

This decoding produced steady consistent results as a digital stream. The SSE lab converted the digital output to audio for us to play here at the Café."

"Sounds like packet radio," I offered. "Yeah, it might be," agreed Felix.

We both listened for a minute or so, along with the other Café patrons.

Felix said, "You know, this sure does sound like some radio transmission here on Earth that has become mixed with the signals from the star, via reflection, cross-talk, power line pickup, whatever. But the SETI people think they've eliminated those possibilities. They say it is definitely coming from this specific star HD85, and it definitely repeats.

"There's a very good chance that the signal from HD85 is indeed digital packets. We could be listening to ET commercial data going by, or ET space telemetry, or even an ET Internet.

"The signal is definitely centered on HD85. After study of the radio telescope records made it clear that we have been getting the signal about every 54 days, the Parkes observatory in Australia tried pointing their antenna a little away from the star when the signal was due to come in; when they did that, the signal intensity was reduced. Local interference seems ruled out. They tried listening to other stars at the same times as these transmissions are being received, at the same frequency, and they don't pick up this signal. It's unique to this star. If it's a hoax, then it has to be some kind of inside job at several radio observatories."

"So more than one radio observatory is reporting this signal?" I inquired.

"Yes, several are," Felix replied. "The HD85 star is in the constellation Vela, down in the southern hemisphere. What you're hearing now came from the Parkes radio telescope in Australia, but we are getting the same stuff from Mauna Kea in Hawaii and the Atacama Cosmology Telescope in Chile. They're all sending their raw signals here to be run through the SSE."

"What do we know about the star?" I asked. "You called it HD85? I've never heard of it."

After a look at a file on his laptop, Felix began a detailed explanation:

> HD85 is short for HD 85512. The star doesn't have a common name. The number 85512 is the star's index number in the Henry Draper Catalog of Star Spectra.
>
> HD85 is a 7th-magnitude orange star (spectral type K5), 36 light-years distant in the constellation Vela. It was discovered by a planet-hunting team using HARPS (High-Accuracy Radial-velocity Planet Searcher), the super-precise spectrometer mounted on the European Space Organization's 3.6-meter telescope at the La Silla Observatory in Chile. The star's slight radial velocity wobble indicates the presence of a planet which has a mass of at least 3.6 Earths and probably not a great deal heavier than that.
>
> The probable planet is called HD85512b. It is a potentially habitable super-Earth. It could be a water world. Its orbit lies just within its star's habitable zone, which is the region where liquid water could exist given the right conditions.
>
> The planet orbits the star every 54.43 days. It is estimated that temperatures on the planet's surface range from 85 to 120 degrees (F). The spectrum shows plenty of water, so there's probably plenty of humidity. The reason it might not be too hot for life is that the surface temperature of its star is about 3200 K, much cooler than the 5700 K of the Sun."

by Steve Geller

The planet has a mass of at least 3.6 times Earth. There's no way to tell from mass alone whether the planet is small and rocky — like Mercury, Venus, Earth, and Mars — or large and gassy, more like a mini-Neptune.

Of course even if the climate of this potential water planet is as balmy as a heated swimming pool, there's no guarantee that it harbors life. Water is the most fundamental ingredient for life, but many other ingredients are also necessary. Because the planet is so far away, it is difficult to observe much more than the mass.

I asked Felix directly. "What do you think we are listening to?"

"I think it is digital data packets," was his prompt reply. "If you listen closely, it's not all repeats of the same stuff. Oh — I did tell you that the stuff playing on the speakers today is a loop, the same thing again and again, right? The message doesn't really come in that often or for that long.

"Some of us think we're hearing leakage from a powerful transmitter, sending commands to a spacecraft. The 54 day interval between messages is because the beam points toward Earth only once per orbit.

"You could be right that it's something like packet radio. I hear a header (that warble) followed by the variable-length data part of the message. That could be a packet."

"Any idea why we haven't heard these messages before?" I inquired.

Felix promptly replied, "We <u>have</u> heard them before. Parkes and the other observatories have been going over their radio archives for past HD85 observations and applying Statistical Signal Enhancement. They found that the HD85 signals had started appearing regularly about three years ago. It might be that the HD85 ETs just started transmitting on this frequency, or their transmitter just acquired whatever feature (or defect) that is causing the signal to leak out to us. Remember, there have been numerous interesting signals received by SETI over the years, but none of them would repeat. This one does, reliably."

Cosmo had turned down the volume on the restaurant speakers. I indicated the patrons still listening on their earphones, and asked Cosmo. "Are they hearing the HD85 messages?"

"Probably not," Cosmo replied. "We just put HD85 on the restaurant speakers for a while today. It's kind of boring. It's annoying some customers, so I turned it down. The people with the earphones are probably nearly all listening to the more usual ET Music, from other stars – or to their personal music collection."

I had to agree that the warble-buzz did get tiresome after a while. But while I listened, I was imagining commercial messages being exchanged on one of HD85's planets, or perhaps weather broadcasts, warning of HD85 hurricanes.

I phoned Dr. Stern at her office. She told me that her students were very excited about the HD85 development. They were looking into all possibilities of packet format. One student was trying various ways to identify the warble part as the packet header; it does seem to be fixed-length.

I mentioned the discussions about telemetry. "That's a good possibility." she said. "Perhaps a spacecraft got launched at HD85 three years ago, and is now exploring their planetary system, like our Voyager probes did here. We're getting leaked telemetry packets from a powerful transmitter used for deep-space communications. That makes more sense than picking up an ET weather FAX."

I thought back to when Pulsars were first detected. At first, nobody believed that such regular signals could be natural; they had to be from Little Green Men. Maybe HD85, or one of its planets, or its local asteroid belt, is in the middle of some natural process that sounds like packets. I tried this idea on Dr. Stern; she said it was too far-fetched.

I asked Dr. Stern how many scientists are buying the notion that ETs from HD85 are sending messages. She said that nearly all the scientists she's talked to recently are waiting for somebody to finally identify an Earth origin for the messages. A few are a little uneasy that we haven't been able to pull out recognizable content. Dr. Stern doesn't agree. If the messages are really coming from somewhere on Earth, even if they are being mangled, we ought to recognize a little bit of familiar code or packet pattern.

She told me: "I'm not sure how we would recognize such a message as for sure extraterrestrial, and not somebody's data packet from Earth. Hey, what would really blow some minds would be to recognize copies of radio signals that we ourselves were sending out some 72 years ago, having made the round-trip to and from HD85. Do you remember the movie 'Contact' in which Earth picked up an ET playback of a 1939 Hitler speech?"

by Steve Geller

From Bits to Bytes

There was another breakthrough from Bob Weiss.

Having identified FSK-encoded bits from HD85, he proceeded to recognize bytes.

Most computer systems today use an 8-bit byte as the smallest unit of information storage in memory; larger data structures are made up of arrays of these bytes. Knowing that the 8-bit byte is the basic building block is a big help to figure out the message structure of a bit stream.

But it appears that whoever/whatever is transmitting from HD85 is not using an 8-bit byte. According to Bob Weiss, the message structure makes sense if one assumes a 12-bit byte.

Going forward from Bob's insight, Dr. Stern's students developed a pretty good case for HD85 to be sending telemetry packets. They have outlined a fixed-length packet header. The same 12-bit byte value marked the start of each header, and each header was 12 12-bit bytes long. Some of the header might have been an address or a checksum. They could see 12-bit and maybe 48-bit numbers in the data part. There was also a possible 5-byte (60 bit) data structure which could be a scaled or floating-point number.

HD85 in the News

The ET messages became big items in the news media. Reporters interviewed both Dr. Stern and Felix.

I tried to do my part for accuracy and completeness in journalism, by emailing some digests of what I'd heard at the Café to reporters at various newspapers. The resulting news stories mostly got things right. Here's an example:

> **Message from the Stars**
>
> *SETI has finally picked up a message from an ET.*
>
> *Newly developed computer equipment at UC Berkeley has processed signals from a radio telescope and extracted what appear to be actual message packets from an ET civilization. The packets are coming from a star called "HD85" which is 36 light-years distant from Earth.*
>
> *Dr. Ruth Stern, head of the UC Berkeley Statistical Enhancement Lab, says the probable planet sending the packets has an average surface temperature between 85 and 120 degrees (F). The spectrum shows plenty of water, so there's probably plenty of humidity.*
>
> *Another UC astronomer, Dr. Felix Fanchot, said "Because the planet is so far away, it is difficult to observe much more than the mass. There's no way to tell from mass alone whether the planet is small and rocky — like Mercury, Venus, Earth, and Mars — or large and gassy, more like a mini-Neptune.*

by Steve Geller

But water is the most fundamental ingredient for life. And now this far-off warm moist world is sending us messages. Stay tuned, Earthlings.

Notice that the newspaper called Felix "Dr." – he wouldn't get his PhD for another couple years. Maybe he was writing his thesis about ET messages?

Later on, yet another newspaper speculated that the HD85 messages, when decoded, will turn out to be an ad for vacation condos on the tropical islands of this warm moist world. The newspaper did not suggest how interstellar condo purchasers would get to HD85.

To my mind, the most perceptive comment in any paper was a quote from a teenager. He said that the messages weren't intended for us, and that picking them out of the radio telescope signals was like hacking into an email exchange between people who were using a foreign language.

Of course, some of the tabloid papers tried to make the HD85 packets sound like either a sign of an imminent alien invasion or religious revelation about the end of the world.

I ran into Father Paul at the SETI Café and asked him if he'd read the papers. He shrugged off any thought of theological content in HD85 messages. "If it's really anything extraterrestrial," he told me, "it's far more likely to be a weather report or a shipping schedule for one of HD's businesses. Actually, anything that tells us about the local culture of HD would be most interesting. And we'd be far more likely to understand a business communication than religious preaching." I had to agree.

I made a call to Karen Banks in the linguistics department. She said that nobody in her group had been able to make any language sense from the HD85 messages. She agreed with me that they were telemetry – arrays of numbers.

Nobody among the UC researchers has been able to make any detailed sense of the HD85 messages, but the simple fact that they've been picked up has stimulated efforts to detect messages from other stars, using their new tool, the SSE.

by Steve Geller

Orange Juice from an Orange Star

Cosmo had done his bit for HD85 publicity. There was a new item on the SETI Café menu:

> *Orange Juice from an Orange Star*
> *Fresh-squeezed Orange Juice, from California oranges.*
> ***Inspired by*** *HD 85512, a solitary orange, star approximately 36 light-years distant in the constellation Vela. This star is approximately one billion years older than the Sun. HD85512 is known to host one low-mass planet, named HD 85512b , orbiting just inside the habitable zone. The planet could potentially be cool enough to host liquid water if the planet exhibits more than 50% cloud coverage. HD 85512b is currently the third best candidate for habitability according to the Habitable Exoplanets Catalog.*

Perhaps that was an excessively verbose description. I was told that when Abner, the cook, saw this description, he shouted "Hey, it's just orange juice!" I think either Felix or Dr. Stern must have handed Cosmo an article from one of the technical journals.

I was sure this "water planet" would turn out to be something hot and poisonous, like Venus, and the HD85 message was actually coming from some place closer to Earth. I mentioned this idea to one of the newspaper reporters, but he didn't use it.

After the HD85 media stories stopped appearing, most SETI Café patrons lost interest. The messages were not at all musical. They were in fact boring, and there didn't seem to ever be anything new. A few people claimed to channel something from the warble-buzz. One guy said it made him think of fried eggs.

Abner must have overheard that remark because he decided to add "Eggs HD" to the menu. It was a variant of Eggs Benedict using a plant-based egg substitute, some polenta instead of ham slices and something other than Hollandaise sauce; not sure what it was. Eggs HD sold quite well at first.

by Steve Geller

Dot-Dash

A clever suggestion from a Cal undergraduate caused a surprising discovery.

He convinced a friend who works with the SSE software to try a very simple filter on the signals from stars. Instead of looking for packet formats, FM radio or other such complex schemes, this filter looked for regular gaps in the signal, dots and dashes, the kind of primitive signaling which was used in the early days of wireless radio communication here on Earth.

Success was promptly achieved in processing radio signals coming from the star Tau Ceti, picked up by the local radio dish located in the hills west of Stanford University.

The signal at one frequency showed a short burst of a dot-dash pattern, which looked like Morse code. Gaps had been appearing in this and other signals, but they were thought to be artifacts of SSE processing. With the gap filter, the code pattern became evident.

The code pattern lasted for just 15 seconds at a time, and appeared irregularly over a period of several weeks. It was the same pattern every time. It's as though some ET Marconi on a planet of Tau Ceti had been trying out his newly-built transmitter, and the test signal happened to leak out toward Earth.

The dot-dash signal seemed too good. One would expect a "primitive" radio transmitter not to be narrow-band, but to spread over a band of frequencies. But the carrier of these dots and dashes was very much narrow-band –showing up only very close to one specific frequency.

Of course it wasn't actually Morse code, but the dot-dash pattern sure looked like it ought to be. A little more software processing identified distinct different groups of dots and dashes, separated by gaps; the groups presumably represented letters – at least that would be the case with Morse code sent on Earth.

Soon, a list of the dot-dash code groups of "Tau Morse" was worked out.

To make the message easy to look at, each code group was assigned to a printable letter. The most frequent code group was assigned to 'E' and so on, following the frequency of occurrence of letters in the English alphabet. This produced the text string NSAXBUTPOADPNFAIFSFA .

The student spread the news about Tau Morse and Tau Text. Very soon, patrons at the SETI Café were listening to the code and studying the text. Tau Ceti is about 12 light-years distant. No other radio telescopes had picked up the dot-dash at the same time as the Stanford dish. So-far, it looked like nobody else was listening to Tau Ceti at the right time, or if they were, they were not listening on the right frequency.

The Tau Text message did not made any sense. For fun, the code groups were converted from letters to musical notes of various pitches and durations. This generated more "ET Music" for Café patrons to groove on.

by Steve Geller

I stopped in at the SETI Café with my laptop, ordered some Tau Tofu, and listened to the raw dots and dashes for a while. It does sound like telegraphy, but rather slow and deliberate. I used to be able to read International Morse Code, but none of this Tau stuff made any sense to me. I tried the music version. It was pleasant tinkles, but nothing more. The text NSAXBUTPOADPNFAIFSFA didn't mean anything to me whatever I did to it.

I had a conversation with a regular Café patron named Frank, who seems to know something about ancient languages. When I described the Tau Text as gibberish, Frank told me about the "Voynich Manuscript," which consists of cryptic character sequences, so-far undecoded. The manuscript was acquired in 1912 by a Polish book dealer. It has been Carbon-dated to the early 15th century.

There are several books about the Voynich manuscript. Maybe the text is actual gibberish, deliberately composed by someone as a joke. It could also be a code, but it seems strange that in all these years, there has never been anyone able to read it. Also, the manuscript has plenty of fancy lettering and pictures; I've looked it up on the Web (just Google "Voynich Manuscript"); it's quite a work of art.

Frank then went on to tell me that if English text were all run together without spaces between words, it might still be readable by someone who was familiar with the language, but to anyone else it would be gibberish. He said that Roman Latin around 200 AD was written "scriptio continua" — without spaces between words or even spaces to separate sentences, because in those days paper was expensive and reading was an elite skill. The elite were supposed to be so familiar with the writing in use that they could recognize words even without the spaces.

Frank pointed out that scriptio continua remains in use today, in Internet addresses like "www.newmediacampaigns.com"

While I was pondering the Tau Text on my screen, my bearded friend Starman looked over my shoulder. He offered an interesting suggestion: the letter 'A' appears quite regularly; maybe it marks the space between words. Possibly, but the "message" still made no sense to me, in any language.

I'm sure glad I was born into a time and place when literacy is nearly universal, so none of us have to deal with foolishness like scriptio continua. But I still can't make any sense of Tau Text.

The SSE lab kept processing radio signal files from Tau Ceti, but there was nothing new to report for over two weeks. Maybe the ET with the code key had switched to a different radio frequency.

by Steve Geller

Morse from another Source

Then a civilization at a second star began tapping a code key.

The Atacama Cosmology Radio Telescope in Chile reported a burst of dot-dash in a radio signal from HD 10180, another Sun-like star with a large planet close-in. Like the signals from Tau Ceti, the transmissions came as dots and dashes at one frequency and appeared at irregular times. The pattern was always the same, but different from the Tau Ceti pattern.

The probable planet is a large gas giant, about the size of Saturn, with an orbital period of 5-6 days. There is probably an Earth-sized planet there too, but unlikely to be life-bearing because it's too close-in and hot. The star HD 10180 is 127 light-years distant in the southern constellation Hydrus.

The new source was provisionally labeled "HD10."

The HD10 code groups were also identified and assigned to letters. So-far, nobody has been able to get any intelligence from the resulting text. There has been only the one transmission to work with; nothing else has arrived since.

Now that two "beginner" ET transmitters had appeared, the researchers began looking for more signals with dot-dash patterns. It made sense that there should be a lot of eager young ETs just getting started with radio.

HD28

During those days, the SETI Café during the afternoon sometimes resembled a study hall, with people sitting at the tables pondering possible messages from the stars. Most had a laptop computer in front of them, earphones on their head, and a plate of star-inspired food next to the computer. They were either studying printed representations of HD85 packets or trying to make sense of Tau Morse or just listening to the usual ET Music. At least they were eating while they did these things, and the Café was making money.

Someone had set up an "ET BLOG" which purported to be an "Alien Internet." It was mostly sponsored ads, but it did give some flavor of what it might be like to have an ET Internet connected to ours. Message headers had fake email addresses like "zorbar@bigmoon.epsilon.eridani.et" and "Marconi@tau.ceti.et" Somebody pointed out that the "et" domain is reserved for the country of Ethiopia, but the Alien Internet bloggers kept on using it.

One day, during this time, Dr. Stern, Cosmo and I got together for an afternoon meeting. I had some Tau Tofu while Cosmo ate Eggs HD. Cosmo remarked that he really likes the Eggs HD, but sales of it have fallen off, possibly because people think it is too fattening. It used egg substitute and polenta, but Abner's "Hollandaise" sauce turned out to be cheese based.

Dr. Stern announced that she was going to order some Goldilocks Oatmeal, because it was appropriate for what she had to tell us about the latest star message discovery. I picked up a menu and soon found the item she was talking about. I read it aloud:

by Steve Geller

Goldilocks Oatmeal
Oatmeal with Fruit and Nuts
Inspired by *the star HD 28185, a yellow dwarf star similar to the Sun about 138 light-years distant in constellation Eridanus. It has a planet which takes 1.04 years to orbit its parent star. This orbit lies entirely within its star's habitable zone, also called the "goldilocks zone," where conditions are "just right" for Earth life. But the planet is big, like Jupiter, and may not have a solid surface.*

"Right," said Dr. Stern after my reading. "What's amazing is that when I wrote that astronomical description for the oatmeal menu item, I had no idea we'd actually hear something from that star. I thought the Jupiter-sized planet didn't look like a good candidate for ET, but there was a faint possibility we might hear from life on one of the big planet's moons.

"Well, the SSE has now processed a signal from this star, HD 28185. It looks like yet another digital telemetry packet stream, also with bits encoded by FSK, but using a different pair of frequencies from those used by HD85. My students see a header-data packet pattern similar to that which we see from HD85. It even appears to use a 12-bit byte.

"The mechanism for beaming the signal at Earth may also be a fortuitous regular pointing of the deep-space transmitter in our direction, as HD85 appears to do. The telescope Archive showed that they had been seeing about 20 minutes of data about once a year, which makes sense because the probable planet has a period of 1.04 Earth years.

"We're going to give the HD28185 source the short designation of 'HD28.'

"HD28 is 138 light-years distant, in the constellation Eridanus, in roughly the same direction from Earth as Epsilon Eridani. But those two stars are not close together in space. "My guess is that HD28 is yet another ET civilization with a space research program, perhaps operating a Voyager-like spacecraft. They are sending packet telemetry over a deep space radio transmitter, but they don't seem to be using the same floating-point number format as HD85."

I posed a question for anyone in the group of regulars at the SETI Café. "Thanks to Statistical Signal Enhancement, we have picked up what looks like telemetry packets from two different ET civilizations, each of which appears to be running a space exploration program. Why is it that we never pick up ordinary radio broadcasts? Don't these civilizations have news broadcasts, weather forecasts, talk radio? Do they listen to classical music? Is there any ET popular music?"

Felix had joined the group while I was talking, and gave a good answer. "My guess is that an ET deep space transmitter puts out a much stronger – and directional – signal than does ET broadcast radio, so we're much more likely to pick up telemetry than talk radio."

Dr. Stern agreed. "That's true," she said. "Another reason may be that we have been concentrating study on the higher frequencies. ET radio stations may well operate on carrier frequencies closer to our AM and FM bands.

"I think I remember being told that the SSE does not work as well at those lower frequencies. Actually, the HD28 star looks especially likely to have an Earth-sized planet. Maybe we should look at lower frequencies and try harder to detect leakage from their local broadcasts.

"Anyway," Dr. Stern said, "We can be pleased that the SSE has delivered two distinct yet similar alien telemetry streams. The staff of the SSE lab has done a great job. They have great technology, but so much of what they get depends on luck. Running the SSE involves a lot of guesswork. Also, some of the frequencies drift, because of magnetic and electric fields in space. Even when they've gotten locked onto a signal, the situation can change later on.

"I've had several people ask me why the ETs on HD85 and HD28 don't have a SETI program if they support a space program. Why don't we get some messages directed at us? My answer is that they may have some of the same constraints that we do. Their governments may see space exploration as good hard science and see sending messages to ET as expensive playing around. Beaming a message to a possible cosmic neighbor is a high-energy, high-cost proposition. We do very little of that kind of thing ourselves. Why should we expect the ETs to be any different?"

Dr. Stern said that lots of research details will be presented at the up-coming conference on ET messages, to be held the next week on the UC Berkeley campus.

Henry Draper Catalog

Because we kept hearing about stars named HD-something, I finally looked up the Henry Draper Catalogue. This list of stars was constructed as part of a pioneering effort to classify stellar spectra. Henry Draper made the first photograph of a star's spectrum, showing distinct spectral lines, when he photographed Vega in 1872.

The HD catalog numbers are commonly used today as a way of identifying stars. The HD catalog today contains spectroscopic classifications for 359,083 stars.

One of the early contributors to the HD catalog was the Harvard astronomer **Annie Jump Cannon**. It was she, working with E. C. Pickering, who invented the spectral type letters we use today. The original catalog used most of the alphabet. Cannon then dropped all letters except O, B, A, F, G, K, and M, used in that order for giants down to dwarfs. A few more letters were added later to deal with oddball cases. Cannon also defined composite types, for example B5A for a star halfway between types B and A, and F2G for stars one-fifth of the way from F to G, and so forth.

She also subdivided each class using a numeric digit, with 0 being hottest and 9 being coolest. For example, the Sun has the spectral type G2 V, with G2 standing for the second hottest stars of the yellow G class and the V representing a main sequence, or dwarf, star, the typical star for this temperature class.

More than 7% of all stars are G-type. In fact, there are plenty of G2 stars; the Sun is a rather common type of star.

by Steve Geller

A Conference on ET Messages

With ET telemetry messages coming in from two sources, Dr. Stern had organized a scientific conference to discuss all the ET transmissions and their possible interpretation. I attended the main session which was held on the UC Campus. People came from way out of town. The Wheeler Hall auditorium was packed full. I sat in the back row, notebook in hand.

During her introduction, Dr. Stern told all the attendees that the SETI Café in downtown Berkeley was a great place to go for food. I'm sure Cosmo appreciated that plug.

Dr. Stern began by presenting the astronomical information about Tau Ceti.

"Tau Ceti a single-star system. It's a G-class star, similar to the Sun, but only 78% as massive. It is cooler, and its spectrum shows the star to be metal deficient, which may mean it has few if any terrestrial planets.

"No planets have been detected so-far, but there is 10 times as much dust and asteroidal debris around Tau Ceti as there is in our solar system. Tau Ceti has been listed as a target star for SETI searches. It is close, less than 12 light-years distant.

"As you may have heard, one frequency received from Tau Ceti sometimes shows a pattern of dots and dashes, as if someone over there had built a primitive radio transmitter with enough power so that some of its signal leaks out into interstellar space. Tau Ceti is not a variable, pulsating star. I have no theory for a natural process which might produce the dots and dashes.

"I realize that a lot of people are suspicious that the dots and dashes were put into the signal here on Earth. There is definitely a short pause between dot-dash code groups, so everyone is assuming that the code groups represent letters in some alphabet. Several translation schemes have been tried. None of them yielded anything recognizable.

Someone asked if anyone had picked up interstellar Morse code from any other star.

Dr. Stern's answer was: "We have had very short bursts from two sources, Tau Ceti and HD 10180." Then she added, "I'm still wondering about that. I'd think there ought to be more ETs out there who are just starting with radio. Maybe all beginners can't build a powerful enough transmitter. Regardless, I think that more than just the two telescopes should have captured the signals, even if they were of such short duration. Something's fishy. But yes, researchers are definitely listening for dots and dashes from other stars; so-far, no more have been detected."

Dr. Stern then summarized the astronomical information for HD85.

"This star, HD85512, also known as Gliese 370, has a K-type spectrum and is 36 light-years distant, in the southern hemisphere constellation Vela. There is at least one planet, known as HD85512b, which is 3.5 times the mass of Earth and has a year of 54.4 days. We call this kind of planet a 'super Earth' because it seems Earth-like but is somewhat larger and more massive than Earth. It is a rocky planet, maybe with water oceans; it is not a gas giant like Neptune."

One of her students presented the results of packet analysis. The total message length appeared to be fixed at 144 bytes (12 times 12). The transmission rate was very slow – 576 bits/second (48 12-bit bytes per second). Whatever data the ETs were sending must be very valuable. It looked like the same packet was being sent 3 times.

The presenter suggested that the HD85 ETs might have 6 fingers on each hand, based on their evident fixation on the number 12.

That was an interesting observation. I suppose that an ET, observing that Earth packets use an 8-bit byte, might conclude that we have 4 fingers on each hand – or 8. It sure was easy to read too much into this kind of information. The HD85 ETs might not even have hands. But somehow, they managed to operate communications for a space program. My guess was the HD85 ETs do indeed have hands, and not radically different from ours. (But I recognized my human bias.)

By the way, dinosaurs were around far longer than we humans have been so-far, but no group of dinosaurs ever evolved enough intelligence to build cities or develop a technological civilization. One reason probably is that none of the dinosaurs had limbs good enough for grasping and manipulating, or perhaps their overall size was just too big to let them do fine technical work. Pictures of an otherwise fearsome Tyrannosaur show what might be delicate manipulator limbs on the upper part of its body.

I was waiting for a paleontologist to turn up a fossilized knife from a dinosaur site – or a fossilized dinosaur-developed computer!

There was much more interesting stuff presented in the ET Message conference.

There was a presentation from a group of researchers who had been studying the HD85 packet data as numbers.

Within the packets, a 60-bit data unit (5 12-bit bytes) looked like it might represent a floating-point number (an integer plus an exponent used to scale it. This pattern was often seen in the body of HD85 messages.

An older scientist in the audience noted that Control Data Corporation computers (CDC), now defunct, used something very similar to this 60 bit scheme back in the 1960s. He speculated that the souls of some CDC computer engineers may have floated out to HD85 and inspired the alien engineers.

by Steve Geller

An enthusiastic student suggested that we may be hearing a packet stream from the HD85 Internet. He wanted to build a gateway to connect our Internet to the ET Internet and begin surfing HD85 websites. Nice idea. We might have found we needed the services of an HD85 ET interpreter.

Dr. Stern then turned to HD28. She gave a summary of the astronomical information.

"The star HD 28185 is 138 light-years distant. It is a yellow dwarf star, spectral type G5V, very similar to the Sun, which is G2V. It has a planet in the habitable zone with earthlike year and a circular orbit. The planet appears to be big, like Jupiter. It is possible that the HD28 packets are coming either from a different, undetected, Earth-sized planet, or from one of the giant planet's moons."

Work on the HD28 packets was still incomplete at the time of the conference. One researcher summarized what had been determined thus far. He said that the HD28 packets were more mysterious than those from HD85. The headers did not always start with the same byte value and the messages were probably of various lengths, but it was hard to tell, because not much of the data stayed the same from message-to-message.

Some 12-bit, and maybe 48-bit numbers were recognized. HD28 telemetry did not appear to use HD85's 60-bit floating point.

These conclusions were stated as tentative. But it did seem very likely that, like the packets from HD85, the HD28 packets represented deep space telemetry.

Someone asked if the ETs at these two stars could be in communication, exchanging technical information. Dr. Stern said this was a possibility. Perhaps the putative SSE at HD28 was good enough to have captured copies of the HD85 telemetry, just as we on Earth had done. But she thought this unlikely, because the distance between HD85 and HD28 is larger than the distance between Earth and HD85. And besides, other than both using FSK, there wasn't that much commonality between the formats of the two stars' telemetry packets.

After the HD85 and HD28 presentations, there was a talk about how the SSE works. This was given by Dr. Thomas Sharp, the tall young black man who was a senior engineer in the SSE group. Here's part of what he said.

"The idea is to acquire a fairly long signal sample that is known to contain non-random data, and then compute statistics to characterize elements of the signal. The statistics are used to rebuild the signal, regenerating elements that were missing, correcting elements that were distorted. It can even suppress regeneration of signal elements which were regarded as interference or noise. Other corrections and enhancements can be applied as part of this processing."

I got the idea that operating the SSE is something of a cut-and-try process. The lab just keeps trying different parameter settings and plug-ins until they get something that is stable and looks good.

by Steve Geller

Dr. Sharp told us that some signals, after enhancement, looked like they might contain lines of information, or an array, which could very well be an image. So-far, this was more guesswork, and not enough information was present to determine an image size, or where it begins and ends. An additional difficulty is how to determine pixel intensity and color. He said they were not sure that the ETs see images as we do, especially colors.

"Do you have any images of an ET?" asked one young student. "No," was the reply, "but some of our people have experimentally jig-sawed together some fragments which might be part of such an image. We don't get any recognizable figure. If we ever get anything close to a complete picture of any ET, we'll be sure to post it on the Internet." That got applause and some chuckles.

I began to wonder if the Voynich Manuscript could be an old record of ET messages, picked up I don't know how in those pre-technology days. As usual, Starman was way ahead of me with an even more exotic interpretation. He thought that the Voynich Manuscript is text left behind by some ETs who visited Earth back around 1200-1300 AD. Asked whether anyone in Europe back then had actually become fluent in the ET language, and if so, should we find more written material in the same script/language as the Voynich Manuscript, Starman had no answers.

News Coverage of the Message Conference

There were at least four reporters at the conference, including one from the Associated Press and one from Reuters.

Here's the story the AP produced, with a little help from me:

Two ET Civilizations have a Space Program

There was a conference held today at University of California, Berkeley, covering the recent discovery of two sources of messages from an ET civilization. SETI is now being taken very seriously.

Both sources are far-away stars. HD85 is 36 light-years away and HD28 is 138 light-years. Both sources appear to be transmitting packet telemetry, the kind of thing NASA has used to communicate with spacecraft like Pioneer and Voyager. Earth is probably receiving these signals by accident; we see them every time the powerful ET transmitter beam, which is presumably pointing at the spacecraft, sweeps past us due to the source planet's orbital motion.

Neither message is meant for us. We are just eavesdropping on whatever part of their far-away science we can figure out. HD85 seems to be using a scaled number format, or "floating point," identical to the one used by a now obsolete family of computers. It's unclear what scaled number format HD28 is using.

But we now for sure have evidence of at least two extraterrestrial civilizations at least as smart as we are, because they're doing their own space research.

by Steve Geller

Because they are so far away from Earth, it doesn't look like either group of ETs will be visiting us soon, but the star around which their planet orbits is, in both cases, very much like the Sun. If they do visit, we might well find that these beings look a lot like us.

One of the UC technical people, Dr. Thomas Sharp, was asked whether the new equipment could pick up a TV program from either of these planets. His reply was "Yes, we probably could, but the signals are going to be very marginal. We'll have to be tuned just right to catch them. So-far, this hasn't happened."

Felix Fanchot, a UC astronomer, said: "We are definitely going to look at several Sun-like stars that are close enough for us to pick up their local radio and TV. Examples are Epsilon Eridani, Alpha Centauri and Tau Ceti. We may soon be hearing an ET announcer."

Someone at the conference mentioned that Tau Ceti is supposed to be putting out Morse Code. Maybe the radio technology of those ETs is still too primitive? Fanchot replied, "Sure, it's possible that the Tau Ceti civilization is just getting started with radio. The Tau Ceti star is somewhat older than the Sun, but evolution might run slower out there. Also, we're not completely sure that the dots and dashes are really coming from Tau Ceti."

When the Stars Began to Speak

Here's another newspaper story, written only about the Tau Ceti dot-dash:

Tau Ceti Key Clicker Just Learning How

The nearby star Tau Ceti appears to be sending us a rather primitive message. The civilization there must be just figuring out how to use radio, and some of their transmissions have leaked out, crossed 12 light-years and were picked up by the big dish at Stanford University.

At one frequency, the signal was broken into segments which appear to be the dots and dashes of an ET version of Morse code. Scientists offer no possibilities for a natural process which might produce the dots and dashes. Tau Ceti is not a variable star.

Code groups were identified, such as dot-dash, dash-dot and dot-dot-dot-dash. These must represent letters in the ET alphabet, but nobody has any idea what they are. To produce readable text, the code groups have been listed by frequency of appearance, and then assigned to letters in the same frequency order as English, in which the most frequent letter is 'E'.

The resulting text is nothing like English. We get strings of letters like NSAXBUTPOADPNFAIFSFA, which doesn't mean much. Language experts at UC Berkeley are working on identifying words in the ET language. Maybe we'll soon realize that the ET has keyed "HELLO EARTH."

Scientists are now listening for other possible ETs with a code key. They may have found one more, in a signal from the star HD 10180, detected by the radio telescope at Atacama, Chile.

Some scientists are suspicious of ET Morse, suspecting that somebody has interfered with the recording of radio telescope signals. Other scientists think there should be plenty of such ET radio newbies. New entries to the radio club should be popping up all the time. But it's strange that only two radio telescopes should have captured such signals.

So-far, the messages don't make sense to anyone on Earth, but this may change.

The Dangers of Anticipatory Naming

A reporter noted that the astronomical designations of ET source stars are often too cumbersome and sometimes difficult to pronounce, so when an interesting new source is discovered, it gets a short and simple name.

He asked me why the SETI researchers use HD85, Tau Ceti and Epsilon Eridani and other such cryptic names for stars. Why can't we use something more descriptive like "water world" or "packet-sending star"?

I was with Felix when this question was asked. He gave a short lecture about the dangers of anticipatory naming in science. His remark later appeared as a sidebar in one of the news articles. I can't find the article, but here is a synopsis.

For an example, Felix cited the detection of "sprites," which are brief flashes of light and other emissions seen above thunderstorms on Earth. The scientist heading one of the first research groups to study this phenomenon didn't want to give it a name which might predispose people to think the physics was understood, such as "electromagnetic surge," so he picked the fanciful name "sprite" because some of the pictures looked like a fairy dancing in the upper atmosphere. Another kind of sprite was designated "elf." These names have stuck and are now used routinely in scientific papers. Felix said that sprites have still not been completely explained, even though at least one satellite has been launched to study sprites from above.

by Steve Geller

Comments from a Conservative

Finally, the radio talk show hosts got on the ET message bandwagon. I heard some remarks by the conservative commentator Lem Rushmore.

Lem strongly suspected that all of the ET messages were fake, produced by scientists trying to promote more funding for themselves.

But Lem also said that if ET civilizations are really out there in great numbers, that fact makes it clear that smart beings like us humans are not going to be destroyed by global warming or running out of oil, so we on Earth don't have to worry about such things.

A Skeptic

At this point, the SETI Café audience encountered a scientist skeptic.

"Hey, here comes one of my colleagues." Dr. Stern beckoned a new arrival over to where she, Starman, Felix, Cosmo and I were sitting in the SETI Café.

She introduced the new arrival to the group. "Please welcome Dr. Samuel Strauss. He teaches Electrical Engineering at CalTech. He was here for the recent conference. Sam is a bit of a critic of our ET messages."

Dr. Strauss was a tall pale man, in his late 60's, mostly bald, with a beaky nose and thick dark-brown eyebrows. He was wearing a suit, which made him stand out from the other SETI Café customers.

"Hello folks," said Dr. Strauss as he sat down. "I hope Ruth is right that there is a proliferation of ETs running space research programs out there among the stars, but I still think all these messages are too much like transmissions from Earth. Some years ago, there was a big flap about ET radio and it turned out that the transmissions were Earth signals reflected back from the Moon. My guess is that either we are seeing Earth messages transformed in some way, or the US Military is going to eventually admit that radio astronomers have stumbled onto one of their classified satellite channels."

I asked, "So you think the alleged ET message packets are enough like packets generated on Earth that they are indeed actually Earth packets, but somehow transformed?"

"That's right," Dr. Strauss replied. "I don't know how the signal gets into the radio telescopes, but one way to generate fake packets is to record packets from one of Earth's shortwave broadcasts, strip off identifying information, then run the packets through a transformation to produce 12-bit bytes."

I felt a bit let-down. "Oh," I said, "so you think the arrival of all these ET messages is a scam?"

Dr. Strauss gave me a grim smile. He replied, "Keep in mind that the search for extraterrestrial intelligence is a fringe activity. It's barely science. It gives wide opportunity for fraud and practical jokes. Remember that SETI heard little or no likely transmissions for a very long time; now we are suddenly getting all these ET messages. -- Yes, I know about the SSE -- but it still makes me suspicious. It's a fair amount of work to make fake packets the way I just described. I sure wish these jokers would apply their cleverness and effort to useful research."

Dr. Stern asked, "Sam, do you think both the HD85 and HD28 messages are being produced by the same source?"

"Yes, I do," replied Dr. Strauss. "It's very likely. Watch for more so-called 'packet' messages from other stars as time goes by. I can't prove it, but I think somebody has found a way to take a laptop computer to radio telescope sites, hack into the local net and splice the so-called ET signal into what is being recorded from the telescope. This stuff reminds me of the crop circles flap, which was also about messages from so-called aliens."

Felix broke in at this point to repeat his story about the HD85 signal appearing and disappearing as the antenna shifted on and off the source star. "Could this be accomplished by somebody with a laptop?" he asked.

"Sure," replied Dr.Strauss. "All it takes is an inside person. Scientific fraud has been successful from Piltdown Man to Cold Fusion because some people just want to believe exciting stuff. The crop circles were made by people sneaking into farm fields at night. I think we have to look at the possibility of an inside job, collusion going on inside at least two radio observatories."

"You know," Dr. Strauss continued, "those alleged CDC floating point numbers in the HD85 data really bother me. It's easy to brush them off as the work of a fraudster who once worked with CDC computers, but to be totally fair, there's nothing to prevent an ET civilization from independently inventing this or any other computing technology. We can't take the attitude that just because something looks like an Earth artifact, it can't possibly be from an ET."

We were all very quiet after hearing this.

Dr. Strauss paused for a moment, then said "On the other hand, it's all too likely that something that looks like it came from Earth, quite probably did." He gave a short cackle. "Occam's Razor, don't you know?"

Cosmo asked, "What about the Morse code from Tau Ceti? Is that the same joker at work?"

"Probably a different joker," replied Dr. Strauss. "Again, there's nothing to prevent an ET from developing his own telegraph code. And of course there's nothing to prevent an Earth-based fraudster from doing the same thing."

by Steve Geller

Dr. Strauss briefly glanced over at me. "Look, I don't want to pick on journalists, but all the publicity coming out in the news has got to be motivating fraudsters and their copycats. Also, I notice that our code key operator has recently disappeared. Right?"

Dr. Strauss was referring to the fact that the Tau Morse had not been received at all for over a week. When it was coming in, it was broken up with numerous long gaps, some of them lasting a couple days. Also, Tau Morse came in far more frequently than either HD source – and it's always the same pattern. Doesn't Tau Marconi have any other test messages to send?

"OK, Sam," Dr. Stern said. "I'll agree that 'Tau Morse' is probably a practical joke, and the joker has now gone into hiding. But I still hold out hope for the ET reality behind the HD packet messages. Several of my students are trying various ways to make sense of those messages. They are also sending the data all over the world for other students to work on. Something's sure to turn up, even if it is just uncovering a fraud.

"I suppose we should be encouraged by the common features of the HD digital transmissions. They're both based on binary 1 and 0 bits, not on a 3-way code or anything more complicated. It might be that, for all technical civilizations, the simplest scheme is always the best, so that at the lowest level, the same basic technology kit will be developed by everyone.

"Of course, for all we know right now, all the nearby ETs have been exchanging technical notes for years, and we on Earth are just getting to the point where we can join their club."

Trying to Tune In ET Radio and TV

One day, at lunchtime, I was hungry, so I went to the SETI Café and put in an order for the Epsilon Eridani Salad, the healthy dish of chopped broccoli, mixed nuts, carrots and beans in a soy-based sauce. Good food! I can all too easily recall past writing projects that have taken me to places where the available food was much, much worse.

While waiting for my order to be prepared, I stared at the wall poster of the **Hubble Deep Field**. This is an image built up from Hubble Space Telescope long-duration exposures of a "blank" area of the sky in the constellation Ursa Major. The field of view is 5.3 square arc minutes, so small that only a few foreground stars of our Milky Way galaxy lie within it. The area is not blank at all; the long exposure yielded plenty of stars and galaxies. In fact, nearly all the objects in the image are remote galaxies, some showing tiny spirals, some just fuzzy stellar points. It shows how awesomely many stars there are in the universe.

Felix walked in, saw me staring at the wall poster, and joined me at my table. He wanted to tell me some news. He said, "Seeing you stare at the Hubble Deep Field made me think about what I recently heard about the SSE project. I was in their lab yesterday. They are implementing more hardware and software, so they can more easily enhance specific types of signals.

"The original setup worked pretty well for picking up telemetry packets, but now they're tuning for ET radio stations, which might be leaking voice and music into space."

by Steve Geller

"Great idea," I said. "I keep hearing about how radio and TV from Earth has been leaking out into space for years. Any ETs within 80 light-years are supposed to be benefiting from Earth's dubious broadcast cultures, including 'I Love Lucy,' 'The Ascent of Man,' the Vietnam War, our political debates and of course our wonderful advertising. It seems fair that Earth too should be passing through somebody's expanding radio bubble."

Felix cautioned me. "Receiving ET radio or TV really hasn't been possible up to now, because the likely signals have been too weak or too distorted.

"Any AM broadcast using the low frequency band we use on Earth will probably be absorbed by the ET planet's ionosphere, or bounce off; it can't get out, unless their ionosphere develops a hole. Anyway, the lower frequencies lose a lot over the light-years.

"Earth's TV signals would probably be too weak for ETs to produce an image they could look at.

"Also, it's very likely that powerful TV transmitters are a very short term requirement of a civilization. Here on Earth they are being replaced by cable transmission and relatively low power signals beamed down to Earth from satellites.

"The strongest signal currently generated by Earth technology is from the planetary radar on the Arecibo radio telescope; ETs halfway to the center of the galaxy could detect its signal, but we don't run that transmitter very often. It's kind of expensive.

"But the SSE group is trying all kinds of combinations of their settings. Maybe they'll suddenly pick up Epsilon Eridani political talk radio." He grinned.

"The fact is that we on Earth don't put much power into our TV or radio transmissions because the intended audience is not very far away. Our defense radars are much stronger, but still, if ET is doing their SETI research at our level of technology, our leakage signals would be detectable only to the ETs listening within about 155 light-years from us.

"And even if they do hear our carrier signals, the modulation that represents the information content would be much harder to detect, and interpretation would be even harder because we didn't intend the programs to be understood by anybody except other humans.

Still, some clever ET, the equivalent of Bob Weiss, might be able to figure out everything about Earth's radio transmissions.

"It's quite likely that some ET is transmitting radio and TV at about the same power levels we do. But anyway, now, the improvements to the SSE might make it possible to pick up some of the very faint and distorted radio signals that have been beyond our capability thus far.

"The SSE people told me they might possibly have a result from Epsilon Eridani. At just one frequency, they picked up about 25 minutes of carrier with an AM signal. They demodulated the audio and heard some music. There were several 5-10 second bursts of static in it. They think the static might be voice, which might come out clearly if they used the right SSE parameter settings.

by Steve Geller

"The music was plunky twangy stuff. One guy said it reminded him of a Japanese shamisen, a 3-stringed instrument played by both bowing and plucking. Now they are suspicious that they might somehow have picked up a Japanese station on Earth. I guess this is healthy paranoia.

"They also got some FM from both HD85 and HD28, but it was so broken up that they couldn't be sure. They're still working on it."

Fraudsters Found

It was at this point that we found out that all the "Morse code" messages were definitely jokes.

Just as Dr. Strauss had suggested, somebody had imposed the dot-dash pattern onto a recorded radio signal.

The tipoff came when one of Dr. Stern's students tried performing various transformations on the characters of the repeated "Tau Morse" text, which had been generated by assigning the most common code groups to English letters according to their frequency of appearance in English text.

They got a little surprise.

If the string of characters I reported earlier, "NSAXBUTPOADPNFAIFSFA," is shifted back one letter in the English alphabet, and 'A' is converted to a space between words, you get "MR WATSON COME HERE "

Whoops. That's the famous first message spoken over Alexander Graham Bell's newly-invented telephone in 1876.

This discovery immediately raised suspicions and set off an intensive investigation.

A Stanford student soon confessed to having interfered with a recording from their radio telescope to produce the dot-dash pattern. Under threat of prosecution, he implicated a friend who had done the same thing at the Atacama telescope. With some information from the student in Chile, HD10 text was transformed and a different famous message was found: "WHAT HATH GOD WROUGHT."

by Steve Geller

Very funny. OK, we all got fooled. We were not hearing an ET Marconi trying out his code key — not this time.

There was some suspicion of the Cal student who originally suggested processing messages to look for dots and dashes, but he maintained that he was not involved in the telescope data modification fraud. He said he got the idea to look for dots and dashes from someone who knew the guilty Stanford student.

Of course now people began to think the HD packet messages were probably faked too. But nobody confessed. Study of the HD packets continued to be intense and still no fraud give-away patterns were found. Also, no more packet-sending stars had turned up.

After this, the SSE researchers who might have picked up an FM station for Epsilon Eridani became rather closed-mouthed. When Felix visited their lab, they told him not to talk about their results at all. This caution made good sense, to avoid having people to automatically assume a joker is at work any time we picked up an interesting signal.

By the way, "What hath God wrought?" was one of the first messages sent by land-based telegraph, on May 24, 1844. This message was sent just a few miles, over wires, from the US Capitol building to the Baltimore railroad station. Unlike today's SETI researchers, the recipients of that telegraph message probably knew what was coming.

Music from the Void Becomes Better

Several weeks after the dot-dash revelations, I arrived at the SETI Café and sat down at a table. I soon realized that there was beautiful music on the speakers. It sounded like a slow organ piece.

The tables were about half full, normal for mid-afternoon. Most of the customers were staring into space, as if in a trance.

At first, I thought Cosmo had decided to soothe his customers with some comforting organ music, but then I realized that it was some of the Music from the Void that I had listened to earlier. This version was much better. There were no distortions or drop-outs. Evidently the SSE had been put to good use to get better audio.

Most people there looked spacey, and kept listening to the music, as one melodic line segued into another. This cosmic composer sure had done a nice job.

Felix was there in the Cafe. He told me that there was general agreement among the musicians that there were tone parts for at least three and perhaps four instruments: two strings and something else with a buzz in it which might have been a woodwind like an oboe or bassoon. And there was also that drum-like thunk. The soprano part came in sometimes, like a high flute or piccolo; its slow high tone made me think of the aria "Summertime" in Porgy & Bess. There was nothing that sounded like a brass instrument. The woodwind pitch would be sustained for a long period, like a bagpipe drone, then it would fade away, often to be replaced by the rise of a different pitch.

by Steve Geller

The harmony was nice, but it seemed a little unstable to me, as if it were about to clash in discord, but it never does.

I heard a musician telling someone "I've heard those hum tones used in Irish music, to accompany singing. The singer pumps bellows on the tone instrument, and pushes a key when she wants to change the tone note."

A different person spoke up. "I've heard something similar to this in music from the Caucasus, and in Arab music. I don't think this is using Western rules of harmony. I'm not sure I'd know how to write a score for it. Some of the pitches may be between the ones we use in orchestral music."

Felix smiled. "Well, if this is really coming from some far-off invisible star, the rules of music might very well be totally different. Hey, I'd be pleased if it turns out that music is universal enough to be appreciated many light-years away from its source."

One of the Café customers asked Felix, "Have these guys tried the statistical enhancement thing on HD85 or Epsilon Eridani to see if those ETs are doing any music?"

Felix replied, "Yes, they might have detected some music from Epsilon Eridani, but it's not like this.

"By the way," Felix added, speaking to me, "The Stanford People tried pointing their radio telescope to three other 'blank' spots in the sky, and got no signal. It would seem that they were quite lucky with that original blank spot.

"The SSE technology still needs some improvement. It's too easy for the processing to get overwhelmed with too many drop-outs or parts with weak signal. If the SSE doesn't get enough raw materials to generate a steady output signal, it will produce a random rumble or just be silent for a while. It's difficult to tell exactly what's causing that. For this recording, they didn't use the statistics for the elements that they thought were random noise, not music. It's quite possible that they removed too much, perhaps eliminating a voice either speaking or singing."

Felix grinned. "I think there are more interesting things yet to come. The SETI people are still pointing at other stars and trying to pick up stuff like the HD85 packets."

The person who had spoken about writing a score spoke up again. "So-far, I haven't heard any recognizable solos. I don't think there are ever fewer than two instruments playing together, each with a distinct tone. The hum tones shift to make good harmony with the current melody. And that high delicate soprano comes in every so often; I really like that effect." So did I; such lovely cosmic music.

I asked Felix if there was any way to tell how far away in the "void" was the source of the music transmissions.

"We can't know a distance," he replied, "until somebody matches the radio source with a visible astronomical object. There is little if any varying Doppler shift in the carrier frequency, which makes it unlikely that the source is on an orbiting planet."

Felix frowned and added, "Because of the Tau Morse hoax, there's a lot of suspicion of fraud with MV. Observers have been monitoring the operation of the Stanford radio telescope where MV is coming in and have looked for any way for someone to mix this music into the received signal. Everything looked OK.

"For a period of a few days, there was no more lovely music received, but then a new piece came in, starting right during the time that observers were on-site, watching telescope operations. Music from the void does seem to be real."

Someone asked "Why do the ETs broadcast only music on their radio stations? Why don't we get an announcer's voice between musical selections?"

"Good point," replied Felix, "For some reason, all we've picked up is the music. It may be that there's some voice from the void. Every so often the music is interrupted with some kind of interference: intervals of hums, clicks, booms and buzzes which could be a voice. but the SSE doesn't reconstruct anything from these elements.

"I really don't think there's any voice on MV; it seems to be all music."

Felix paused for a moment to think, then he offered a couple of surprising ideas:

"Perhaps MV is not really ET instrumental music. The tones may have a very special informational purpose.

"Even if we're actually hearing radio station leakage from an ET civilization way out there, we might be letting our own biases influence what we hear.

"And it's possible that many of the ET languages simply sound like music to Earth ears. Earth's Chinese and Norwegian languages have a sing-song quality. Hey, it might be that ETs live in an operatic world, singing to each other instead of talking.

"The point is that in any case, it might be difficult to distinguish speech from singing or from music in general."

After I had put in my usual order for Epsiloni salad, I asked Abner if he was going to put a "Music from the Void" dish on the menu. Rather grumpily he replied that the music from the void probably wasn't extraterrestrial, or if it was, it didn't come from a definite place like a planet. He grinned then and added, "I'm waiting for Starman to channel the ET who's sending the void music and then ask that ET to tell us its favorite food." Abner gave a small smile.

Encouraged by the success with securing the music from the void, a radio telescope was pointed to the coordinates of the *Wow!* Signal and several frequencies near 1420 MHz were monitored. The results were run through the SSE.

They got plenty of hiss from Hydrogen all over the target area, but no sharp peak anywhere in the region, at any time. Also, no signal they got had any modulation with music or voice. Whatever made the *Wow!* signal does not seem to be active these days.

When Dr. Stern told me that the SSE group had tried the Wow! coordinates, she added her opinion that Dr. Ehman probably got it right the first time when he suspected that the Wow! signal was the result of a temporary reflection from a piece of orbital debris.

She said, "Maybe it was a large metallic debris chunk, fortuitously shaped like a microwave reflector so that it briefly concentrated some Doppler-shifted Hydrogen background long enough to produce the sharp peak. In any case, I strongly doubt that the *Wow!* signal had anything to do with an ET."

A Signal from Cygni

For its next act, the SSE brought in a signal from the star 61 Cygni. It might have been an ET all-news station, because the audio sounded like voices most of the time.

The 61 Cygni system is 11.4 light-years distant, in the northern constellation Cygnus the swan. It consists of two stars, both dim, just barely visible with the naked eye.

Coincidentally, Cosmo was just about to add a new item inspired by 61 Cygni to the Café menu. Here it is:

Mesklin Meatloaf
Spicy loaf made from soy "meat" with chopped peppers and onions. **Inspired by** *the fast-spinning high-gravity planet Mesklin, said to orbit the star 61 Cygni in Hal Clement's science fiction novel "Mission of Gravity." 61 Cygni is a binary star system, 11.4 light-years distant, in the northern constellation Cygnus the swan. It's dim, just barely visible with the naked eye. It is quite close in the sky to the well-known bright star Deneb, which marks the head of Cygnus the Swan. The 61 Cygni stars are both orange-red, somewhat smaller and cooler than the Sun. So-far, no planets have been detected in orbit about either star.*

In Clement's story, Mesklin orbits one of the stars. The other star is sometimes visible in Mesklin's sky.

by Steve Geller

When I talked with Felix later, he augmented the Café menu description:

> The two orange-red K-type stars orbit each other very far apart, with a mean separation of about double the distance between Pluto and the Sun. The period of the orbit is over 600 years. The stars are far enough apart that they could each have independent planetary systems.
>
> This star pair has a high proper motion, which means that the rate of movement of the star pair within the galaxy results in a large apparent motion of the visible object on Earth's sky, per year.
>
> Because the astronomer Giuseppe Piazzi discovered the fast motion, 61 Cygni has been called "Piazzi's Flying Star."
>
> No planets have been detected, but something is making that radio signal. It could be leakage from local radio on an undetected Earth-like planet – circling either star. The stars are both somewhat older than the Sun.

I thought that the 61 Cygni signal might have been a news or talk station. It was mostly voice, with intervals of music less than a minute long. The Cygni voice sounded like some kind of language. The Cygni music was simple, just one instrument, which could have been a flute or an oboe, playing a rather random melody. It wasn't bad, but I thought it was rather boring.

The Cygni Announcer

The SSE folks worked very hard, re-tuning their equipment again and again to find the best settings to process the 61 Cygni signal, setting the SSE parameters in a way which might especially enhance voice. Finally, they were successful, bringing out clearly what may be the voice of the announcer for the radio program. It definitely sounded like somebody talking, but not with a human voice, let alone in a human language.

Several sequences of Cygni voice were filtered, edited and combined to produce a continuous voice audio, which we called the "Cygni Announcer." This combined sequence was made available on the SETI Café website. Each segment began with about 10 seconds of what was probably Cygni music.

Cosmo, Felix and I had done a lot of listening to the Cygni Announcer. We agreed that most of the voice could have been produced by something like our vocal cords and body structures like our tongue or lips. We heard vowels and consonants. There was a particularly startling high-pitched [*peep!*] which a human might possibly produce when goosed.

Felix again cautioned against letting our human biases make us hear things that aren't there. We might not be actually hearing a voice. But it sure did sound like somebody talking in a totally unknown language with peeps.

There were pauses. But the timbre of the voice always sounded the same, as if the Cygni ETs had only the one announcer. Felix suggested that we may not be able to distinguish a change of speakers, because they all talk the same way. It's also possible that the voice was mechanical, artificially generated.

I could get no meaning from the voice; neither could anyone else at the Café. The linguistics group ran some analysis tools on the audio, trying to generate some statistics for comparing the ET voice with voices speaking Earth-based languages. The group agreed that all the sounds in the Cygni voice, except possibly the [*peep!*], were within the capability of human speech equipment. This raised a lot of suspicion that the voice was indeed a human, putting on an act.

Starman wanted us to play some Cygni Announcer to dolphins, to see if they would react.

The linguists came to a consensus that the Cygni language did not feature tones, like those in Mandarin or Cantonese. It was possible that tones were somehow filtered out when the bursts were processed. There was definitely syllable stress; this Cygni language wasn't monotone like Japanese.

A Spanish Voice from Epsilon Eridani

The engineers at the SSE lab steadily improved their facility. A voice from another ET radio station was picked up. This time, the source was good old reliable Epsilon Eridani, the Sun-like star 10 light-years distant, which was known to have at least one orbiting planet, where salad-eating ETs might live.

We had already gotten music from Epsilon Eridani, featuring the twangy Japanese shamisen. But now a voice had been extracted too. As usual, nobody could get anything from the language. The linguists told us that it was not at all like an Earth language. No surprise. As with 61 Cygni, nearly all the vocal sounds could be at least approximated by our human speech equipment, so in principle, we could learn to speak the Epsiloni language and they could learn to speak one of our languages.

One day, we got an unexpected surprise. Cosmo had brought Pablo out of the kitchen and let him listen to some Epsiloni voice.

"Ay!" Pablo exclaimed, "They're talking Spanish!"

Pablo listened for a while and told us that it sounded like some religious broadcast. Evidently it was good Spanish. Pablo, who came from Mexico, said the speaker was probably from somewhere in South America.

Pablo was delighted by all this. He went back into the kitchen where we could hear him telling Abner all about it.

by Steve Geller

Suspicion of fraud was immediately raised. But there was plenty of Epsiloni language voice there too. The Spanish voice sounded more human than the Epsiloni voice; it was definitely a different speaker.

After listening for a while, Pablo heard a station ID "Radio Santa Rosa from Lima, Peru." A student in the Café recognized that station as a religious broadcaster -- a strong shortwave station.

The Spanish-speaking Epsiloni remained a mystery for a while, but eventually, some investigation determined that the broadcast we had picked up was not recent, and thus unlikely to have contaminated any recent radio telescope data. From internal evidence, the Spanish language program we heard probably was broadcast from Peru about 20 years ago. The Epsiloni language part of what we received might well have been a report to fellow Epsilonis about having picked up a radio transmission from us space aliens on Earth. The report included a playback sample of what had been picked up. Twenty years would be right for a light speed round trip.

Evidently the SSE developed by the Epsilonis did a better job on low frequency AM than ours did, or maybe Radio Santa Rosa once came through at Epsilon Eridani especially strongly, for some reason. Maybe it had leaked through a big temporary hole in Earth's ionosphere.

The linguistics group was now studying the Epsiloni language part of the audio to see if it contained any references to words in the Spanish, which might give some clues for understanding the ET language.

After the Spanish incident, we had Doris Chen, the Café cashier, who is Chinese, listen to some of the ET voices we were getting, to see if any of it was a long-ago Chinese language broadcast. Nothing like that turned up.

by Steve Geller

Common Referents

The initial experience in ET language research highlighted what became known as the "common referent" issue. There is no way to understand a foreign language unless the learner has common referents in both the foreign language and in something he knows – a language or even just a labeled set of pictures.

In the past, when commercial traders traveling to distant parts of Earth encountered a new group of potential customers speaking a strange language, the traders worked out a trade language by getting the words for trade goods and other objects which both trader and customers could use their languages to describe, and build up a "pidgin" language.

In too many of our ET language situations, all we have is the ET language, with no common referents. The repeat of the Spanish broadcast (if that is not yet another practical joke) could provide common referents if the linguists can identify items in the ET language part which refer to things in the Spanish language part.

Hoping to capture yet another ET language, the SSE was fed signals from Tau Ceti, now purged of spurious dots and dashes. They did not find any music or voice – just some noise that one scientist says was caused by random stellar flares. They also tried HD10, once the equally fraudulent "Morse from Another Source." No luck there either.

Talk or Music?

Work continued on music and voice coming from from 61 Cygni and Epsilon Eridani.

It was hard to get much of a start on translating the languages because there was no English language version of any of these ET language messages.

The Spanish from Epsilon Eridani could have been something like a Rosetta Stone, if the linguists could ever find some Epsiloni language voice sequences in which an Epsiloni was talking about something in the Spanish.

Of course, the Epsilonis probably were just as clueless about how to translate Spanish as we were about their language.

In both ET sources, it was not even clear that we're really hearing music or voice. Maybe half of those who have let the music train their minds agreed that the voice was really saying something. Felix remained convinced that for some of these sources, we were not hearing a voice speaking at all, but rather it was all music.

Somebody else suggested that nearly all of what we were calling music might actually be just the way ETs talk. Regardless, nobody seemed to understand what any ET was saying.

by Steve Geller

More News Reports

The news media never made much scandal out of the great Tau Ceti Morse Code fraud. It was mentioned briefly in a few short articles – no big headlines. I think the news people were disappointed. The whole ET Marconi story wasn't quite sensational enough. Maybe some reporters were hoping for a startling translation of Tau Text. Tau Morse eventually dropped off the news.

The Spanish language from Epsilon Eridani was much more interesting. It was big news in the Spanish language press, and some evangelical Christian churches were excited too.

The Spanish language press calmed down quickly when it became clear that the Epsiloni ETs themselves were not actually speaking Spanish. Here is an article that was originally published in an English-language missionary magazine, which is in some way affiliated with Radio Santa Rosa:

> **God Reaches Out to the Stars**
> *Christian radio has been received by intelligent beings more than 10 light-years away.*
>
> *An evangelical outreach Spanish language radio program, broadcast from Peru 20 years ago by Christian station Radio Santa Rosa, has escaped Earth and traveled at the speed of light for 10 years to get to a planet orbiting the star Epsilon Eridani.*
>
> *The Epsiloni beings have signaled that they received the broadcast by repeating some of it back to Earth, another 10-year trip.*

The rest of their return broadcast is in the Epsiloni language, which nobody on Earth understands. Perhaps God intends to teach the Epsilonis some Spanish, or maybe God will enable some Spanish-speaking person on Earth to learn Epsiloni.

What an amazing new missionary field this could be! Several churches are putting together radio programs especially for Epsilon Eridani. They will soon be broadcast from Peru and hopefully will arrive at their cosmic destination 10 years from now. Maybe about 10 years after that we will hear the Epsilonis give their testimony about finding Jesus.

by Steve Geller

One day at the Café, I found a small group engaged in an intense discussion. Seated at one table were Starman, Father Paul and Frank. They were looking at a newspaper article.

They showed it to me. Here's part of it:

Does life exist beyond this planet? Increasingly it appears that Earth has a multitude of brother and sister planets orbiting their own suns.

Sophisticated analyses of data returned by space telescopes may not be able to single out a planet that definitely has life, but it could show that the conditions conducive to life are so prevalent and that promising planets are so numerous, that the likelihood of plenty of life out there becomes overwhelming.

At a recent conference held by NASA and the Library of Congress, historians, philosophers, and theologians gathered to ask how humanity should prepare for the discovery of life beyond planet Earth.

The conference addressed the challenge of moving beyond current conceptions of what constitutes life, intelligence and civilization — conceptions which are now based on anthropocentric models.

One panel discussed the philosophical and theological implications of a universe potentially teeming with life.

Some participants argued that finding even primitive microbial life elsewhere would severely shake up humanity's way of thinking about itself. It might create a watershed akin to Galileo proposing that Earth was not at the center of the universe.

Some even saw it as a death blow to religion.

It was that last line which had provoked the intense discussion.

As I arrived, Father Paul had begun to give an interesting response:

"The concept that life can only exist on Earth limits an infinite God. The more things we discover, the more wonders of God we will find.

"The Bible has many references to a "God of heaven and earth." Psalm 147 says: "He determines the number of the stars and calls them each by name."

"This surely indicates that His creation includes much more than just our planet, more beings than just us.

"One could say, in fact, that claiming anything less than a universe brimming with life is an insult to the glory of God."

Father Paul gave a big grin.

Well, good, I thought. I guess finding ET doesn't have to mean the end of religion.

by Steve Geller

Amateurs Transcribe the Cygni Announcer

A small team of amateur researchers (Starman, Cosmo, Abner and I) attempted to transcribe the Cygni Announcer and produce a phonetic written script. We played a 1-minute section of Cygni Announcer again and again, slowing it down and speeding it up, in an attempt to write down what it was saying, using a consistent notation.

We couldn't agree on what we had heard, so we made some reasonable compromises. Here is a short segment of what we produced:

> *dnookh dookh vahlah [peep!] gsdsa'a psu dnehk nnnoog [peep!] ahlah [peep!] uahka dookhk*

?Starman translated this somehow as "Listen up people, the ETs are talking to you."

I showed this transcription effort to Karen Banks of the linguistics group. She gently told me that our transcription was too crude, and that the linguists have better ways to do the job. I saw one of their transcriptions and it looked just as incomprehensible as ours. Well, whatever. Nobody still knew what the ETs were saying.

Our transcription was useful in one entertaining way. Some of the patrons of the SETI Café now greeted their friends by saying "peep dookh." (Starman likes to think this is saying "how're ya doin' dude." The response was usually "nnnoog peep ahlah peep." (doin' fine, and you?).

A very few people were able to produce a quick whistle for that [peep!]. Everybody else just emitted a short scream – eek!

I remember thinking then that if this pseudo-ET chatter ever goes out into space as part of an Earth radio program, someday some SETI researchers at 61 Cygni are going to be most puzzled – if they don't begin laughing hysterically.

by Steve Geller

An Epsiloni Tries to Speak Spanish

Pablo Pino kept listening to the recording of voice from Epsilon Eridani which featured the Spanish from Radio Santa Rosa. He found an interesting section, and had me listen to it and produce a transcription.

This is a portion of what I got:

> *Durka f'lala pen ghiza skhal pengwo.*
>
> *Ay hoona frendeh day ezpranza ee ayoodha prah steh ee prah mee oon leebo.*

The first line is probably straight Epsiloni ET language, but the second line Pablo and I think is the Epsiloni attempting to quote some Spanish, but he has a strong Epsiloni accent.

We think the Epsiloni is talking about something his fellow Epsilonis received from an ET source (us on Earth), but neither Pablo nor I have any idea of what comment is being made.

Using my transcription, Pablo tried to filter out the "Epsiloni accent." He rendered the second line in standard written Spanish as:

> *Hay una fuente de esperanza y ayuda para usted y para mi, un libro*

He translated this into English as:

> *There is a source of hope and help for you and me, a book*

Given the known Christian broadcast source, the book was probably the Bible; we're not sure what the Epsilonis thought it was.

Pablo did not find any other Epsiloni Spanish quotes.

by Steve Geller

Starman the Zombie

Starman made an interesting claim. He had been fascinated with the music from 61 Cygni. One day, after a few hours of listening, he began to have an experience of vague images in his mind, like those reported by the people who first listened to the original ET Music in the Café. He began to get strange ideas, almost like memories, as if he were recalling stories somebody had told him long ago.

He began to listen to Cygni music and voice together. After a while, he thought he could understand a little of what the Cygni voice was saying.

Starman said he felt as if an extraterrestrial personality had taken residence in his mind and was guiding him in interpreting the voice of the Cygni Announcer. He wanted to call having his own resident ET "being a zombie." I thought he was using the term incorrectly. Starman convinced several friends to listen to the music and try to repeat his zombie experience, and soon there were four other zombies all giving translations of what the Cygni Announcer was saying. The zombies weren't doing a literal translation, but rather producing a rough paraphrase of what was said.

I tried using the music to train my mind, but the Cygni Announcer still made no sense to me. I guess I don't provide a good home for an ET personality.

I asked Starman to explain what he thought was going on.

"I think the music is modifying my mind," he told me. "I think that's why the ETs are sending the music. The music somehow sets up my neural circuitry so my mind is receptive to learning the ET spoken language."

"The Cygni Announcer makes sense to you now?" I asked.

"Well, sort of," he replied hesitantly, "I haven't learned to understand the ET language the way I understand English, or even like the way I partially understand Spanish. Maybe I would achieve that after a while. I just get a vague sense of what is being said. I may recognize a few words or phrases, but mostly I just come away with a feeling that I know what the voice intended to say."

"So what is the Cygni Announcer saying?" I inquired.

"It might be some kind of religious sermon," Starman said. "It calls for promoting cooperation, learning from one another. The idea is to reduce tension and increase appreciation for what we have in common as well as ways groups of people can form a larger more capable whole person. We're being told that we need to cooperate, help each other, to learn from one another."

Several of the other zombies were listening to my conversation with Starman. They all agreed that Starman's summary was the substance of what the Cygni said. Starman looked a little embarrassed. He asked me, "Please don't report this as revelation from 61 Cygni. I'm afraid the Café will fill up with religious fanatics. I want to make it clear that we are not hearing an actual sermon. We just get the feeling that this is what the voice is saying."

I said "You mentioned being able to understand a few words. What's an example?"

by Steve Geller

Starman picked up a copy of the Cygni Announcer transcription we'd produced. He said, "I think the word that sounds like 'nnnoog [peep!]' might be the Cygni word for 'help'. Another sound 'dookh' might mean something like 'you' or 'people'. We hear 'dookh' a lot.

I can't say we know enough words to string together to make sentences. We're just guessing about 'nnnoog [peep!]' and 'dookh' because we hear these words when we get the idea that the Announcer is saying something like "you should help other people."

While I was talking to the zombie group, Felix arrived, pulled out a chair for himself, and gave the group a report on his recent activities.

"I saturated myself with the 61 Cygni music," he said. "Then I played the Cygni voice, and I generally agree with what Starman and the other guys tell us is being said. But this was after I'd talked with them and heard their impressions.

"There's another possibility. Both the music and the voice may somehow be evoking such thoughts and ideas in our minds, so that we think the voice is actually saying this stuff. My point is that we might not be dealing with language in the same sense as English, but we're still experiencing communication. I definitely like the idea that the music is being deliberately sent to train our brains."

"You think both music and voice are aimed at us?" I asked.

Felix shook his head slowly. "I doubt that," he said. "I think it's much more likely that we are the fortuitous recipients of leakage from an especially powerful broadcast transmitter on 61Cygni, designed to give wide coverage on their planet. It has the unintended side-effect of generating a narrow beam which sometimes points our way — something like those religious short wave radio stations that take advantage of ionospheric bounce late at night to get wide geographical coverage. It's probably a broadcast intended for local consumption on the ET's home planet. The music might be intended to facilitate religious emotion among the locals, and to some degree it works on us Earthlings as well."

At this point, we were joined by Father Paul Beni.

Father Paul asked, "Did you say the ETs are sending religious broadcasts?"

We all must have looked uncomfortable. Starman answered the question. "Maybe, we're not sure. We think we might have heard something inspiring from an ET, which could be a sermon."

"Could I listen?" Father Paul asked.

Starman handed over his earphones, and gave Father Paul a quick summary about the Cygni Announcer and the idea that the Cygni music teaches our brains how to comprehend the words. He let Father Paul listen to the Cygni music for a while, then played some of the Cygni Announcer. Father Paul listened intently for a few minutes, then slowly took off the earphones and handed them back to Starman.

"The music was kind of like singing," he said. "The other stuff might have been words, but I sure couldn't get any sermon out of it." He grinned and we all chuckled.

by Steve Geller

Starman suggested that Father Paul listen to the music for a long while, at least an hour, then try the words again. He explained how to access the website for Cygni data and find cleaned-up recordings of both the music and the voice. Father Paul thanked us and said that he would go home and try listening on his own computer.

I was a little worried about Father Paul. I could see the possibility that he, or members of his congregation, might get the idea that there was divine revelation to be heard from 61 Cygni. Felix is probably right that, with or without the music, a person is likely to hear what they want to hear in the voice.

Mormons not surprised by ET

After Father Paul had left the Café, Starman told me about talking with a Mormon, who said that the leaders of his LDS Church were not surprised that we were hearing from all these ETs.

The Earth's creation, according to Mormon scripture, was out of existing matter. The Mormons say that this Earth is just one of many inhabited worlds, and that there are many governing heavenly bodies, including a planet or star they call "Kolob" which is said to be nearest to the throne of God.

This Mormon wanted to know if anyone could get ET messages from Kolob. Starman told the Mormon he'd ask around.

by Steve Geller

Visions from the Void

My favorite ET music had always been MV -- "Music from the Void." It was so beautiful, and there was no voice in it.

I recorded one especially affecting MV piece, a long series of rising and falling flute-like tones mixed with two bass lines, lasting about 12 minutes. I then played it for myself, at home, several times. I found the result interesting -- and a little frightening.

I fell asleep while listening to the MV recording and had a vivid dream, of floating among brilliantly lit clouds. I wasn't quite asleep. It was more like a waking vision.

I had some difficulty re-connecting to reality after the music ended. I almost didn't want to come back. The experience was very absorbing and almost thrilling. I emerged feeling refreshed and upbeat. After about a minute with no MV music, I returned to normal awareness.

The experience was more than a little frightening. If the MV recording had been on a continuous loop, I'm not sure I would have been able to shut it off. Is MV some kind of ET addictive opium?

The next day, I went to the SETI Café with the intention of comparing notes with Starman, Cosmo or anyone else who might have had my experience.

When I arrived, a disturbance was going on. A middle-aged woman was yelling at a young man who was sitting at a table. The young man had been wearing earphones for listening to his computer; he was staring vacantly, with tears running down his face.

The woman had pulled his earphones off and was waving them in front of his face. "Wake up!" she kept shouting. As I approached the table, the young man began to stir, emerging from his reverie. He shook his head, rubbed his eyes and said "I'm OK, Mom." His mother grabbed some paper towels and began to wipe off his face. The young man was smiling.

"I'm OK, OK." he repeated. "Oh, that was so beautiful."

I blurted out, "What were you doing? Were you listening to Music from the Void?" "Yeah," he replied. "How did you know?"

His mother grabbed my arm, and demanded of me, "Do you know what it was Jim was listening to? We live just down the street. Cosmo called me when Jim began to cry and they couldn't snap him out of his trance."

Jim, now mostly restored to reality and emotional stability, began to explain.

"Yes, I was playing one of the 'Music from the Void' selections from the website. It was really beautiful. I kept playing it again and again. I guess I fell asleep and had a dream. I met the spirit of my little brother Eddie. We were floating up in the sky, surrounded by bright clouds. Eddie looked so happy."

Jim's mother broke in to tell the gathered crowd, "Eddie died last year. He drowned when he hit his head while diving. Jim and Eddie were very close. What music was it that affected Jim so deeply?"

Cosmo and some of the others explained about the MV source and how it was obtained.

Jim said, "I never was affected like that before, but I guess I never kept playing that music so many times before."

I announced loudly: "I recently had a similar experience."

I told the group about my dream of floating in the clouds. "For me, the experience was all beauty, a great vision. I didn't encounter any spirits, but I do remember seeing numerous blobs of light going in and out of the clouds."

"I gotta try this!" That was Starman. "Which selection did you play?" He looked at me, then at Jim.

"You know, I might have gone through the same thing." This was Frank, one of the Café regulars. "For me, it was like dreaming of encountering great people of the past, like Gutenberg, Copernicus, Michaelangelo. It happened late at night. I must have fallen asleep while the MV was playing. My dream was very vivid."

Cosmo looked alarmed. "Whoa, guys!" he said. "We may have found something dangerous about ET music. I don't want anyone to get hurt."

"OK," said Starman with a grin. "Let's give it a check. I'll try buzzing myself with some MV. Somebody get ready to pull the plug, or at least take my earphones off. Say, Cosmo, didn't you play the MV for quite a while on the speaker system that first day? Did any customers zonk out?"

Cosmo looked alarmed at first, then quickly replied, "I don't remember any problems. Some people got kind of dreamy-eyed. Nobody cried. Nobody had to be slapped back to reality." He shrugged.

"I'm going to try this," Starman declared. "Jim, show me what file you were playing."

Jim showed him, then Starman asked two customers to sit by him, one to run the music, the other to pull off his earphones. The audience had expectant, apprehensive looks. Cosmo did not look pleased, but he didn't interfere. Neither did I.

"This is nice stuff," Starman said as the music began. He leaned back in his chair and closed his eyes. "I'm not getting any visions yet, just a feeling of peace. Ahhhh… Hey! You need to keep clicking to keep the music going."

His assistant/protector clicked again, and again …. Starman relaxed.

"No visions… uh … wow!" He fell silent. His other assistant was about to grab his earphones, but Starman waved him off. "Gosh, this is beautiful…. Give me about a minute more, then shut it off."

by Steve Geller

Everyone waited. Starman looked relaxed and happy. The first assistant stopped clicking so the music would stop. About a minute later, Starman opened his eyes, shook his shaggy locks and gave his audience a big toothy grin. We all relaxed. Cosmo, who may have been about to call 911, took his hand off his telephone.

"Better than Weed, man. Better than LSD, even…. Cosmo, don't look at me like that! It's really not that far out; just nice. The Feds aren't going to come raid your restaurant."

"What did you see?" I finally asked.

Starman described his vision. "It was like being out in space. Everything was black, with stars twinkling here and there. I could see a big round ball below me, a planet covered with clouds. The sun was behind me, I suppose; the clouds were brightly lit. As I watched, two little moons set behind the planet, one after another."

"I asked, "Did you have any emotional reaction?"

"Well, happiness, I suppose. I didn't cry, did I? Hey, I'd do this again, but maybe it would be a good idea to have someone with me, or an automatic shutoff of the music."

I definitely agreed with those precautions.

An intense discussion then began among the Café customers, everyone offering ideas about what had really happened, and how the music had caused it to happen. Nothing was agreed.

Cosmo said he was going to password-protect the MV files. I don't think he ever did.

Beings on Big Planets

I was studying some new posters on the back wall of the SETI Café.

One showed the swirls and bands of the atmosphere of Jupiter. Just left of center, above a white band, a little ball was visible, suspended above the cloud tops; this was the satellite Io, famous for its volcanism. The orange lava fields were barely noticeable at this scale. Io is a little larger than Earth's moon.

Huge Jupiter, in the background, looked like a vast banded wall.

Starman saw me and beckoned me to sit at his table, where he was having a discussion with Felix and Frank.

"You know," said Starman, "We may be missing some big stuff. The extrasolar planet searches keep turning up big planets, similar to our Jupiter, Saturn, Uranus and Neptune – the gas giants. These worlds are easier to find because they are big and massive; but they may be at least as common as small rocky planets like our Earth. Maybe some of our ET messages are coming from big planets? Could an ET live on Jupiter?"

Starman certainly could be depended on to bring up interesting ideas.

Felix proceeded to give us a short lecture:

> The gas giant planets are not thought to have a solid surface where rivers and oceans could form and people could live and walk about.
>
> Jupiter's atmosphere is mostly Hydrogen and Helium, the primary substances of the Sun. Jupiter is almost a star itself. In fact, if Jupiter were about 80 times more massive than it is, then nuclear reactions would begin in its core and the Sun would become part of a double star system.
>
> There are other components in Jupiter's atmosphere, including relatively small amounts of Methane, Ammonia, Nitrogen and Oxygen. There's Sulfur and Phosphorus too. These trace components give Jupiter's banded clouds their distinctive coloring.
>
> Water is thought to be present deep in Jupiter's atmosphere, but the directly measured concentration of water at the top of the clouds is very low.
>
> The atmosphere of Jupiter probably does not have a clear lower boundary. Deep down, with the increasing pressure, the gases gradually compress into liquid. This liquid layer is thought to sit on top of a solid core of metallic Hydrogen and Helium. Jupiter does not really have a solid surface like that of Earth, Venus or Mars.
>
> The same is true of Saturn. Uranus and Neptune are similar but may have more water mixed with their Hydrogen and Helium.

If there's Earth-like life in the vicinity of the giant planets, it is probably going to reside on one of their moons. Jupiter's Europa and Saturn's Enceladus both appear to have liquid water oceans below their ice surface.

There has been much past speculation about life floating high in the atmosphere of Jupiter. It might be ammonia-based life. Sagan and Salpeter wrote a paper in which they raised the possibility of life in the upper regions of Jupiter's atmosphere. The oceans of Earth have simple photosynthetic plankton at the top level of the food chain, fish at lower levels feeding on these creatures, and then marine predators which hunt the fish. Sagan and Salpeter envisaged airborne creatures floating in Jupiter's atmosphere, feeding on ammonia-based krill. These creatures would be giant gas-bags that move by pumping out helium. They calculated that some of these "Jovian Whales" might grow to be many kilometers across, possibly visible from space.

Arthur C. Clarke wrote a science fiction story about such gas bag beings. It was titled "A Meeting with Medusa."

Felix called our attention to the posters on the back wall of the Café. The one next to Jupiter showed an artist's notion of Jovian Whales. They looked like stubby blimps, with huge eyes on top and small fins to the side and rear of their body. Several are shown floating above yellowish roiling Jovian clouds. In front of the body, just below the eyes, is a gaping mouth, presumably used for inhaling the Jovian analogue of krill.

by Steve Geller

Felix said, "There are plenty of extrasolar big planets. There might be plenty of life on big planets, but just not Earth-type life.

"I have a file here on my computer about the star Gliese 876.

"This is a red dwarf star about 15 light-years distant. There appear to be four large planets orbiting the star. The two middle planets are similar to Jupiter, while the closest-in planet is thought to be similar to a small Neptune or a large terrestrial planet. The outer planet has a mass similar to that of Uranus. The orbits of all but the closest planet are locked in a rare three-body Laplace resonance."

"Resonance?" I inquired.

"Yeah, " Felix began another explanation:

> Resonance means that the orbital periods of three planets are related by a simple ratio. Jupiter's moons Ganymede, Europa and Io are in a 1:2:4 orbital resonance. That means for every 4 orbits Io makes around Jupiter, Europa makes exactly 2 orbits and Ganymede makes exactly one.
>
> Gliese 876 planets E, B and C have orbits in that same ratio." He consulted the file again. "That's periods of 30, 61.1 and 124.3 days.
>
> I just found out that several radio telescopes have picked up regular burst noise from Gliese 876. The pattern of the noise matches the motions of those resonant planets.

> I can imagine Sagan's Jovian Whales having evolved to use electric sparks to generate radio calls, like Earth's whales do with sound.
>
> We definitely receive electromagnetic bursts from Jupiter, and to a lesser degree from Saturn.
>
> The longer bursts are caused by charged particles spiraling in Jupiter's strong magnetic field. Shorter bursts appear to be the result of lightning strokes in Jupiter's atmosphere. There are also bursts with downward swoops in frequency; the audio version sounds to me like cries.
>
> The atmosphere of Jupiter is very active. Jupiter gets energy from the Sun, but the big planet also has internal energy sources. There are a lot of storms. There's a lot of violent upwelling – just the kind of thing that produces lightning in the atmosphere of Earth.

Felix gestured at the image of Jupiter and Io behind him and said, "In that picture, some of the swirls are larger than the Earth. Jupiter is an awesomely huge world.

Most of his audience was enthralled by the notion that we could hear from some life form flying around in a planet's huge atmosphere, chirping at radio wavelengths, perhaps by making electrical sparks? Electric eels can make electricity. Bats use high-frequency sound. Whales make low-frequency calls.

Songs of the mile-long Jovian Whales – wow!

by Steve Geller

Starman really liked the idea of gigantic gas bag whales, crackling their electric calls as the whales floated in the vast roiling atmosphere of a giant planet.

He thought that life on gas giant worlds could well be more common than Earth-like life on rocky planets. The Jovian Whales might live among swarms of similar life forms of various sizes, in a thick gas giant atmosphere just as fish live among swarms of krill in the currents of Earth's oceans.

Starman and Frank began talking up the idea that tribes of intelligent gas bags are talking to one another over long distances as whales do on Earth.

I could visualize whales roaming Earth's oceans, using an evolved language which enables them to coordinate migration and operate a social system. I could believe that some of the songs of the humpbacked whales are recitations of traditional tales. The Jovian Whales may well do similar things.

However, given that we have achieved only a very sketchy understanding of the conversations among Earth's whales or dolphins, I see little prospect of understanding conversations among beings like Jovian Whales, especially those floating in the atmosphere of far-away gas giant exoplanets.

Are there ETs Close to Home?

Inspired by the talk about Jovian Whales, I did some reading about animal communications.

Here on Earth, some kinds of animals use sound signals to coordinate action and pass information to others of their kind. Humpbacked whales, in particular, sing songs which are heard over vast distances in the oceans.

Only we humans have languages well-developed enough for transmission of tradition, publishing of news and writing of science fiction stories.

Other animals have language capability, but we humans see their complexity of communication as of a far lower order. Birds and whales can tell other members of their species where they are and perhaps where food is to be found.

As far as we know, only humans have a writing system – unless we've failed to recognize some markings on rocks or trees as an animal's written record. I suppose the scent markings trailed by ants aren't quite writing, nor are the scents left by dogs on trees and light poles, although these inscriptions are clearly being read by the species of animal that wrote them.

Some of the accounts I read suggest that whales and dolphins are calling to each other. An individual dolphin has a "signature whistle," which serves as his name. Other whistles may be used to enhance dolphin group cohesion, something like shouts of encouragement. We do not, so-far, have the ability to produce a translation of dolphin chatter or whale songs.

by Steve Geller

Dolphins, suitably motivated, might have the ability to learn to speak a human language. One experimenter got a dolphin to ask for a ball to play with by producing the sound "bawww." This dolphin even could call the experimenter, whose name was Margaret, by making the sound "Maaahwgwit."

Some types of dinosaurs may have had voices and could call to each other across Cretaceous canyons. The fossil skull of a Parasaurolophus has a bony tubular crest that extends back from the top of its head. The crest is shaped like a trombone; it might have been used to produce distinctive sounds, probably low rumbles which change in pitch.

We already know that human beings use over 6000 languages, and that they are very different -- English, German, French and Spanish are quite different from Chinese, Vietnamese, Korean and Japanese.

Starman and some others were sure that the voices from Epsilion Eridani and 61 Cygni were all just as complete language systems as American English.

If we were living as gas bag entities communicating by crackling lightning high in the atmosphere of a gas giant planet, that crackle language would be easy for us to understand because it was ours. To gas-bag beings, American English would be random incomprehensible noise.

One day at the Café, I was swapping some short-wave listening stories with a fellow patron. We'd both had the experience of hearing a broadcast in a totally unknown language.

I told him my story of riding a campus bus along with participants in a Berkeley mathematical conference. I heard many languages. I could recognize some of them: German, French, Spanish and Russian. Mathematics seemed to be a common language; there were numerous exchanges of sheets of notebook paper, covered with cryptic symbols. Two people sitting just in front of me were conversing in a totally unfamiliar language; I couldn't get any of it, except for an occasional English reference to something in Berkeley. I later found out that I had been hearing Turkish.

No amount of listening to a foreign language will produce any comprehension – unless there are some clues to establish what is being communicated. The Cygni Announcer voice is incomprehensible. We have no context. Without context, any language appears to be gibberish.

I had just remarked that gas bag beings on Jupiter or Gliese 876 were, like whales, of a lower communications order than humans, when someone munching their Tau Tofu at the next table pointed out that some humans endlessly repeat nonsense on talk radio and in political speeches, yet we treat this as intelligent communication.

OK, I stood corrected. Communication is strongly cultural. One must be inside the culture to communicate in a way that is more than rudimentary. We here on Earth are so far outside of the ET cultures that there is no communication at all. There's barely any communication with Earth whales.

A few SETI Café patrons had been listening to audio recordings of radio noise from Jupiter, Saturn and the Gliese 876 system. It's probably rather boring, but maybe these people liked the idea of listening to the calls of extraterrestrial whales. So-far, nobody has claimed to have channeled ET thoughts from them.

by Steve Geller

A biologist I talked to at the Café liked the idea of gas bag beings on giant planets. Like other people, she noted that some fish on Earth are able to produce electric discharges; maybe lightning sized discharges can be used as a means of communication if the being is big enough.

She pointed out that bird calls on Earth convey simple signals like "I am here, and of this species." "I'm looking for a mate." "There are predators in the vicinity." Howler monkeys are thought to use similar signaling.

Sending a Message to ET

Dr. Stern had been out of town; I hadn't seen her for a while, so when I came into the SETI Café, I was pleased to see her there talking with Felix and two people I didn't know. The group was having a lively discussion in subdued tones.

They seemed occupied with their discussion, so I didn't butt in. I noticed they were all eating the Epsilon Eridani salad. I put in an order of it for myself and sat down with my laptop to check my email.

The subdued conversation went on for about five more minutes, then I heard somebody say, "Hey, we could ask our local science writer!" and I was drawn into their discussion. They wanted me to tell them, historically, how many deliberate messages Earth has sent to the stars, and specifically whether any one of those messages was constructed so that the ETs would recognize it as originating from an intelligent civilization – and be able to understand what was sent.

I consulted some files I had on my laptop and gave them the following summary:

by Steve Geller

The first such message was sent from the big Arecibo dish in 1974. It consisted of an array of 1679 bits, and was aimed at M13, the great globular star cluster in Hercules. Any ETs who picked it up were assumed to be smart enough to see that 1679 is the product of two prime numbers, 23 and 73. The ETs would then use that information to display the bits as a 23x73 grid of dots. If they did this, they would see a few simple pictures. I guess an immediate reply wasn't expected because M13 is 25,000 light-years distant, but there sure are a lot of stars in that cluster – more than 30,000 – and thus a lot of possible ETs.

Straight analog audio has also been sent. In 2001, a radio telescope in the Ukraine was used to send electronic music, the wailing sounds of a Theremin, which is an entirely electronic instrument played by moving one's hands near "antennas," which are connected to capacitor plates. The Theremin music was beamed toward six nearby stars. We may be receiving something like that kind of message now, but MV doesn't seem to be a Theremin. We're still not sure how universal music really is.

There have been numerous other efforts to beam Earth's favorite music out to several different stars. It's not clear whether our music would sound intelligent to an ET. We are quite sure that we recognize musical intelligence in the hum tones we hear from the void. It has been pointed out many times that leakage of our radio and TV has been providing abundant evidence to anyone in the universe who can listen, of the existence of technology-competent cultures here on Earth. Some people thought this could be dangerous, because we may be inviting invaders.

The first serious effort to send a "here we are, and we're smart!" message went out in 2008 from Ukraine's radar to the red dwarf star Gliese 581. The signal was something of a digital time capsule, with music and voice contributions from numerous celebrities. No effort was made to train the ETs to understand it; they were on their own.

In 2009, the 70-meter antenna at Tinbinbilla, Australia was used to send text messages to the 'd' planet of Gliese 581. A website had collected goodwill messages from the public for 13 days in August 2009. Gliese 581d is one of the exoplanets thought to be able to support life. Its distance is 20.5 light-years.

I had not heard of any efforts to construct a message that's sure to be both recognized and understood. Nobody seems to know just what is required to accomplish this. Sending prime numbers or Fibonacci numbers probably would signal some level of intelligence, but our languages still would not be understood. And here on Earth, I was not aware that whales had reacted in any way when they listened to a list of prime numbers. But apparently some whales showed appreciation of a clarinet solo by Dave Rothenberg.

Several science fiction scenarios have imagined the aliens beginning their message with a copy of something they had received from us. The message from Epsilon Eridani with Spanish in it could be an attempt by an ET to communicate with Earth, but it is more likely to be random leakage from an ET local news report describing how their scientists had received "ET" radio.

by Steve Geller

Felix spoke up then, to suggest some general principles for communicating with an ET civilization. "First, we have to attract attention by sending something the ETs are sure to recognize. Then we have to send something which could only come from an intelligent civilization. Maybe we can assume that mathematics is universal. Or maybe the common language at this first stage is not mathematics, but music."

Those ideas made me think about Starman's zombie notion of music being used to train our brains to understand the ET language. Could we construct some kind of music to train ETs to understand English? The idea still seems far-fetched. If we're going to communicate with any extraterrestrial, we and ET simply have to share some common referents. This will always be nearly impossible when we're separated by all those light-years.

Later on, I came across a quote from the biologist Lewis Thomas, author of "The Medusa and the Snail." He said:

> This language [music] may be the best we have for explaining what we are like to others in space, with least ambiguity.
>
> I would vote for Bach, all of Bach, streamed out into space, over and over again.
>
> We would be bragging of course, but it is surely excusable to put the best possible face on at the beginning of such an acquaintance. We can tell the harder truths later.

Visiting the SSE Lab

I finally arranged a visit to the Statistical Signal Enhancement (SSE) lab on the Campus. This is where signals from the stars are enhanced so that we can listen to music and voice.

My visit was set up by Felix Fanchot and Dr. Stern, both of whom came along with me and some others, to listen to the presentation.

After their success with so many signals, the SSE lab funding had been increased. It was a sizeable operation the day I visited. I saw at least four separate work stations, each of which had an equipment rack with the hardware part of the SSE plus a computer which ran the software part and connected to data servers. At one of the stations, the equipment rack was open and a technician was doing something to the plug-in circuit cards inside it.

I greeted Dr. Sharp, the SSE senior engineer who had spoken at the ET message conference. He showed us around.

Dr. Sharp said that they process signals by first setting up the SSE hardware, which involves plugging in some small circuit cards and setting switches. Then they start the SSE operating software and import radio telescope files from a data server. The next step is computing statistics. This process takes much longer than any signal duration. The statistics are then used to control the regeneration, how an output signal is produced.

by Steve Geller

One of the SSE techniques is to use signal segments of various lengths to compute the statistics. Sometimes, for no clear reason, using some segments in this way produces much better output results than using other segments. There's a lot of guesswork involved. Dr. Sharp called this "heuristic tuning."

I asked how much processing of the signal was done at the radio telescope.

"They do a lot," was the reply. "They digitize it, remove jitter produced by the telescope hardware, and compensate for the frequency shift produced by the orbital motion of the Earth. They also compensate for the space motion between us and the source star, which is in its own orbit within the galaxy. For example, the whole 61 Cygni system is moving toward us at about 64 km/sec; that produces a strong Doppler – radio waves shifted toward higher frequencies.

"Here in the SSE lab, we try to compensate for the source planet's motion around its star, if we know which planet is the signal source."

Dr. Sharp indicated a different looking work station nearby. "Here is where we do post-processing. When we've extracted a signal, the hardware here demodulates the audio, filters and smooths it to reduce algorithm-generated noise."

I asked, "Do you apply any motion compensation to the Music from the Void signal?"

Dr. Sharp shook his head. "MV has no varying Doppler, which we expect from something in orbit. We can't measure radial velocity, because we don't know what the un-shifted carrier frequency is supposed to be.

"Our post-processor can pull out audio from any signal. It can deal with any analog or digital modulation that we know about. After Bob Weiss detected FSK in HD85, we made sure we can deal with all variations of that too. We try stuff until something works.

"There have been dropouts within the MV music, which may be the result of source motion disturbing the SSE lock. I don't really know. I just know we've somehow pulled out this lovely music.

It's still very much a possibility that we've been contaminated by somebody's local radio transmitter, but right now, I sure don't see how."

I was slightly confused.

Then Dr. Sharp proceeded to confuse me some more. "Here's another little mystery," he continued, showing us a printed copy of an image. "The optical telescope at Lick Observatory stared hard at our MV blank spot and produced this little smudge. I'm told that it is a remote galaxy, and that this galaxy might be acting as a gravitational lens, magnifying the signal intensity from an object behind the galaxy. Lick got a rough spectrum of the galaxy smudge, and it looks like it has a red shift corresponding to a distance of 8 million light-years. If it's really a lens, then the music source has to be much farther away, which is very long ago in look-back time. If so, the music-making ETs would have to be an awesomely ancient civilization. They might not even be there anymore, and we are just receiving their residual performances."

I must have looked stunned. Dr. Sharp grinned at me. "Well, it still could be that the MV source has nothing to do with that smudge galaxy; it may just happen to be in the field of view. Maybe there's an MV source star quite close, but especially dim for some reason."

We all left the SSE lab then, Felix and Dr. Stern went to their offices, I went for some comfort food downtown at the SETI Café. Thinking about the SSE gave me a headache.

A Summary of the Situation

The SETI Café had become a popular gathering spot for scientists, students, journalists and the general public who are following developments in messages from extraterrestrials. A visitor could always find somebody to talk to about ETs. The level of expertise varied, but the level of interest and enthusiasm was always high.

The star-inspired food continued to sell well. The best seller was the Epsilon Eridani Broccoli and Nut Salad, closely followed by the Mesklin Meatloaf and Tau Tofu; the latter remained quite popular in spite of the Tau Morse fiasco.

Another big seller was Abner's recently-developed Jupiter-themed dish. One of the customers had suggested the Café should start offering Jovian Whale Steak, but this was quickly vetoed as politically incorrect — a sizeable number of SETI Café patrons were members of Greenpeace -- and besides, the Café was supposed to be vegetarian. Abner settled on "Jupiter Red Spot Yogurt" which is white yogurt stirred with colored sauces to produce stripes, blobs and bands; there's a little red berry to represent the Great Red Spot. It does kind of look like Jupiter. For some customers, Abner mixes in a few chunks of cookie, representing the Jovian Whales. I was much impressed. Grumpy Abner could be quite imaginative.

Abner had already created "Eggs HD," inspired by HD85 and HD28, but he refused to be inspired enough by the Music from the Void to create a dish for it. He maintained that MV was not from an ancient ET civilization but rather some natural cosmic singing we have yet to understand. He scoffed at the heavenly visions.

by Steve Geller

The original artificial ET Music was still being produced from several stars, and some Café customers kept listening to it, but there was much less talk of channeling ET, probably because most people saw the original ET Music as contrived and artificial, while the packets, music and voices were really from ETs.

The Music from the Void kept coming in. The cosmic composer didn't seem to run out of musical ideas. Still, nobody knew who/what was making the music. No star had been identified as the MV source, and it was very hard to accept that the musicians really were way out there in the realm of the galaxies, and aeons ago in look-back time.

SETI research had become extremely popular at UC Berkeley. There were plenty of students working in Dr. Stern's group, and in the linguistics group. Both groups were conducting regular seminars in which people reported progress on their various projects.

The students studying HD85 and HD28 were frustrated, because they couldn't make much sense of the data in the packets they were studying. Since they didn't know what was being measured, it was difficult to say much about this data. One number in a packet might be a distance from an orbital center, because it increases and decreases around an average value over long time periods. Since they didn't know what distance units were being used, none of them could say whether that average value was close to the orbital radius of any of the likely planets. The only consensus was that they were receiving a signal probably not beamed deliberately at Earth, and that the signal was probably part of some deep-space communications project local to the remote star system.

by Steve Geller

Do the ETs have a Meter Stick?

Can we expect the ETs to base their units of length on something universal? It can't be the length of the human foot or the standard Meter stick. How about the speed of light?

I checked Wikipedia, and found this:

> *The metre (meter in the US), symbol m, is the base unit of length in the International System of Units (SI). Originally intended to be one ten-millionth of the distance from the Earth's equator to the North Pole (at sea level), its definition has been periodically refined to reflect growing knowledge of metrology. Since 1983, it is defined as the length of the path traveled by light in vacuum in 1 / 299,792,458 of a second.*

That article did not encourage me. I found it hard to believe that any ET would just happen to adopt something close to Earth measurement units. Someone at HD85 might well choose a unit of length based on the speed of light in vacuum, but would they define their "meter" unit with the exact multiple 299,792,458? Not unless they had been using our meter stick.

Interstellar measurement standards? Not likely. Hey, here on Earth, we can't even standardize on the spelling of "meter."

The other numbers in the HD85 packets might be 12-bit numbers. Some might be characters coded in a 12-bit byte, or pairs of 6-bit characters.

When I explained the 60-bit floating-point number possibility to Starman, he suggested that since they came up with something that turned to be the same as what we had, the HD85 folks might actually think a lot like we do, which could mean that they are using an alphabet not wildly different from ours.

That would be so nice. We could be sitting here reading the HD85 Wikipedia and getting the answers to all our questions.

by Steve Geller

Who Are We Hearing From So-Far?

So we were now receiving messages generated by ET life of some sort in orbit around the stars Epsilon Eridani, 61 Cygni, HD85, and HD28. We knew how far away these ETs are -- none closer than 10 light-years. That was way too far away for conversational message exchange. The possible gas-bag beings on Jupiter were the closest ETs we thought we knew about.

Also, we had yet to receive any written messages. The "Morse code" was a joke. HD85 and HD28 appear to be sending us just numbers.

We might not even recognize an alien alphabet. ETs might not use an alphabet.

What About MV?

How close is MV -- "Music from the Void?" Could MV be from a star close by, but too dim to detect? Possibly, but Lick Observatory did detect that fuzzy faint far-off galaxy near the MV position. If that galaxy was really a gravitational lens that is amplifying the MV signal, then the MV music source must be very far away indeed – millions of light-years. Felix thought the music CD title should be changed from "Music from the Void" to "Music from the Beginning of Time" – because it might be coming to us from an ancient look-back time.

I mentioned this look-back time issue to Cosmo and he suggested that if MV is really that far away, then the music cannot originate from any ET civilization at all, because it comes from a time before any intelligent civilizations anywhere had evolved. I hadn't thought of that.

Cosmo also mentioned another idea I hadn't thought of. If the universe is "closed," meaning that it does not go on to infinity, but bends back around, MV could be music from our own very distant past. So why don't we recognize any of the melodies?

Starman, as usual, had a clever answer: the music is coming from a previous Earth civilization, one which was wiped out long before what we call the Stone Age. One of his friends suggested that maybe the great civilization was a group of ancient ET visitors. Yow. Speculation was getting way ahead of the facts again.

One musician friend of Starman's still was convinced that at least some MV music was an obscure musical composition by John Cage.

Conflicts and Rebukes

The extensive publicity about the ET messages finally attracted the attention of some government and religious authorities.

Father Paul Beni was rebuked by the local Catholic Bishop, for mentioning during a mass the probable substance of the rather vague sermon we thought we'd received from the Cygni Announcer. Father Paul was told to stop doing that. I guess the Bishop wasn't sure the Cygni Announcer was Catholic. One Berkeley parishioner had publically speculated about Jesus visiting ET planets. The Bishop discouraged that kind of talk too, because the idea has no Biblical basis. Jesus certainly didn't mention ETs in his parables and we don't hear about any of the saints having an encounter with an ET.

A group of local Mormons had asked Dr. Stern if SETI could get messages from Kolob. Dr.Stern told them that they'd first need to come up with the coordinates in the sky for Kolob. This was a stopper. Evidently Mormon scripture is rather vague about the location of the Throne of God and the allegedly nearby star/planet called Kolob. When the group passed this rebuff back to LDS authorities in Salt Lake City, the word came down to cease such inquiries and stick to reading the Book of Mormon.

Several leaders of various faiths around the world expressed concern that people were being fooled by the ET messages. They said they were worried about unauthorized interpretations some people might be making from the partial translations of ET messages. Already, some fringe people had claimed they had found "new revelations."

by Steve Geller

One Muslim cleric reminded his followers that Allah had delivered a final revelation through Mohammed, and that the Koran contains no mention of messages from ETs. Mohammed is said to have made a "night journey" covering the 766 miles from Mecca to Jerusalem, which also included a leg up to heaven and a return to Mecca by morning. Some Muslims had been rebuked for suggesting that Mohammed made some side-trips to ET planets. I guess the Islamic authorities are OK with ET being out there on other planets, but they want to keep Prophet Mohammed closely tied to our Earth.

A few members of the US Congress objected to any communicating with ETs, claiming that we could be encouraging an alien invasion. Were they worried about ET immigrants? I think these Congress people finally realized that SETI was only receiving messages from the stars, not sending messages out.

A Russian linguist claimed to have identified a few Russian language forms in the language of the 61 Cygni Announcer, and maybe even a few distorted Russian words too. He wanted to send a Russian language broadcast to 61 Cygni to see what happens.

Someone in Israel was sure he heard some Hebrew words from the 61 Cygni Announcer.

There would be more such claims. People sure do hear what they want to hear.

Conservative Casts Doubts

Conservative commentator Lem Rushmore continued to cast doubt on the reality of any ETs.

He said that if the beings making the messages are really technologically intelligent like us, then they ought to be enough like us that there should not be all the difficulty over language. He quoted some scientists who said that what we think is language and music is a processing artifact, caused by our super fancy SSE trying too hard to extract an ET message.

Cosmo, Felix and I agreed not to publicize any accounts of visions experienced by people who listened extensively to MV. We wanted to wait for more analysis. Both Cosmo and I were worried about a possible reaction from the controlled substances people in the US Government, if they got the idea that the SETI Café was pushing an addictive ET thrill pill. We were also concerned that Lem Rushmore might spread some dangerous distortions if he heard about the MV visions.

by Steve Geller

Still Thinking about ET TV

The TV issue kept coming up in conversations at the Café. People asked: Why don't any ET planets leak their television or Internet? That's still a good question. It was clear that TV from Earth leaks out to the stars.

I had been exchanging emails with Bob Weiss, the clever retired engineer, keeping him up to date. Finally, Bob sent me a nice summary of the TV technical issues. Here's part of it:

> *I think it is pretty easy to recognize analog TV, e.g. NTSC and PAL. We earthlings have gotten away from analog TV, and now use digital, which is a stream of packets making up a `transport stream'. A transport stream usually carries several `program streams' each identified with a `program ID' [PID]. In some respects decoding a transport stream is similar to decoding an Internet stream.*
>
> *One difference though, is that digital TV is always compressed, e.g., mpeg-4. This is not encryption, but it might be difficult for ET to figure out.*
>
> *Digital TV reception is sort of like FM reception. As the intensity of the signal weakens, there is no change in the received signal until it crosses minimum signal strength, then nothing is received.*
>
> *Another possible form of modulation is `phase rotation modulation'. This would be a `linear-polarized' radio wave that is rotated back and forth (less than half a turn) for the modulation. This can be done very rapidly using electronic means.*

Clouds with free electrons between us and the source will rotate the polarization, but it will appear as a DC offset in the received signal.

In addition to rotating the polarization, the electrons also retard the signal. This is what limits the accuracy of GPS — unless you compensate for the effect by using two harmonically related frequencies in the same path.

From what Bob said, then, it looked like receiving ET TV might be as complicated as understanding ET languages. We do, however, have some common concepts: the pixel (intensity and color), line and frame. Of course, it's possible that the ETs have been using some entirely different technical scheme for making a picture into a radio signal.

Maybe some ETs have been able to decode some of our old analog TV. If they have, they sure haven't yet played any of it back to us.

by Steve Geller

Another Conference

I was invited to attend Dr. Stern's second conference on the ET messages, held on the UC Berkeley Campus. The group was small this time, about 30, mostly local campus people. There were presentations from astronomers, linguists and the SSE group.

The astronomers went over the sources of the messages and the likelihood of their being from a planet with a technological civilization. The SSE people described the types of signals they have processed, and their reliability and quality. The linguists described the ET languages, and their efforts to understand them.

Dr. Stern pointed out that most of our sources were probably planets similar to Earth, small, rocky worlds with water and a UV-shielding atmosphere. It's possible that some civilizations have developed on moons orbiting large Jupiter-like planets, but the only moon we know of with an atmosphere is Saturn's Titan, with its cold liquid-hydrocarbon lakes.

She summarized: we believe that we have really received ET messages from the following sources:

61 Cygni -- 11.4 light-years distant; pair of orange dwarfs; no known planets.

Epsilon Eridani -- 10 light-years distant, a sun-like star. It has a planet with 1.55 Jupiter mass, two asteroid belts; there could be some rocky Earth-like planets. The message may be leakage from a planetary FM station. There's both music and voice. Somebody there can play a shamisen.

HD85 (HD 85512) -- 36 light-years distant, sun-like star. It has a planet with 4 times Earth mass. Messages appear to be telemetry packets, possibly from a deep space transmitter used for HD85 space research.

HD28 (HD 28185) -- 138 light-years distant, a yellow dwarf star similar to our Sun. It has a planet in the habitable zone with earthlike year, but the planet is big, like Jupiter, and may not have a surface. We could be hearing from ETs living on one of the giant planet's moons. Messages appear to be telemetry packets, similar to but not the same as those from HD85.

Dr. Thomas Sharp, from the SSE group, tried again to explain the technology behind the SSE. Most of it still went way over my head. Dr. Sharp said the SSE is working well, especially after the staff developed easy and rapid methods to adjust its parameters. He said that SSE is still very experimental and still requires some guesswork and fiddling to get good results.

by Steve Geller

Some people at the Café had expressed suspicion that SSE output has little to with its input, but rather is mostly produced by some unintended side-effect of its internal operation. Someone asked Dr. Sharp if he was sure they were really extracting cosmic signals, not producing internal artifacts. Dr. Sharp acknowledged that, because the SSE is so complicated, this possibility is a valid concern. He said that the SSE group frequently repeats the processing of a set of data, using different parameter settings. Also, the radio telescopes routinely deliver several data sets, some from pointing the telescope slightly away from the supposed source.

Dr. Sharp then spoke briefly about "ET Marconi." He said that we should not discount the possibility that we might pick up leakage from a civilization that had just begun to run a radio transmitter. We might indeed hear the dots and dashes of an ET Marconi with a code key. True, a primitive low-power transmitter would probably not leak into space, but some ET civilization might get lucky and somehow develop a crude but very powerful transmitter, and that signal would leak out.

The point was that we should not stop looking for a "Morse source" just because somebody faked one from Tau Ceti. Code groups could be a great way to capture an ET language in a written form.

The linguists were up next. They played recordings of ET voice received from 61 Cygni and from Epsilon Eridani. They showed transcriptions they had done and said that neither language was like any one of Earth's. Neither ET language has tones. The Cygni and Epsiloni languages did not seem to be related to each other.

The linguist doing the presenting at this point was a blind man named Dr. Louis Engel. He remarked that being blind had helped him concentrate on the ET music and voice. He said that he himself hadn't gotten much out of any of the voices and that perhaps the zombies were kidding themselves.

Dr. Engel said that the linguists had contacted speakers of many different Earth languages and played the ET voices for them. Nobody recognized anything (except for the Russian and the Israeli). As far as the linguists can tell, the languages are truly extraterrestrial.

Someone asked about the Spanish language voice segment received from Epsilon Eridani. Dr. Engel said it was definitely Spanish, and consistent with other broadcasts from Radio Santa Rosa. Personally, he was suspicious that somebody had worked the Spanish into the radio telescope data. The Spanish broadcast does appear to be from 20 years back, but it's possible somebody could have got hold of such a recording and fed it to the telescope.

Dr. Engel said the linguists were very much intrigued by the small effort made by Pablo Pino and me to find a piece of ET language which refers to something in Spanish. They've been working on it intensely, but the fragment was too small to build much of a common referents list.

Dr. Engel then went into some detail about the common referents issue. He said that translating ET languages may be impossible unless we get more of an idea of what they are talking about – some common referents. He said he wished that at least one of those "I was captured by a UFO" stories really indicated that we are having ET visitors; he would like to invite a few of those UFO visitors to stop by the linguistics lab and give him some help. He'd even provide parking for their UFO. That got a laugh.

by Steve Geller

Then he got serious. He described how good writing, especially technical writing, starts with getting words and concepts clearly defined before using them. The teaching of languages works the same way. Any attempt to create a "universal language" will have to be based on this principle. We might get lucky and encounter an ET who has gotten a start on a universal language, maybe even a group of ETs who have already begun universal communication. Dr. Engel offered his opinion that the spacecraft telemetry that we have been receiving from HD85 and HD28 might be a significant start.

Amateur Linguists and LEX

The last linguist to speak at the conference was Karen Banks, the graduate student who had talked with people at the SETI Café.

She told us about clubs of amateur language learners, all over the world. They are a brainchild of an organization known as the Institute for Language Experience, Experiment & Exchange, also known as LEX, which was created in 1981 by Yo Sakakibara in Japan, where the clubs are especially active today.

In LEX club activities, both children and adults acquire multiple languages simultaneously, using natural language acquisition – the way children learn a language.

Club members get together at least once a week for interactive activities. They listen to speeches, conversations and songs, in different languages. Those members who feel confident, may recite back from recordings, even if the sounds are not accurate, and even if they don't exactly know what they are saying; the goal is simply to get a feel for the language sounds and tones of voice. Members play games, sing songs, dance, and speak with each other in the various languages they are learning.

by Steve Geller

Karen said that she belongs to one of these LEX clubs, and had acquired partial fluency in 5 different (Earth) languages. She said she had heard that a few LEX clubs in California have been studying the languages of the ET voice messages. They are hampered by the lack of common referents, let alone an English translation, but they absorb what they hear anyway. Some club members have become rather good at mimicking the 61 Cygni Announcer, using approximations for the difficult sounds. One group in San Diego is particularly interested in the voice from Epsilon Eridani, because that civilization is "close" to Earth, and because of the Spanish in it. A few club members say they can almost feel like they're talking about the Spanish broadcast when they are mimicking the Epsiloni voice.

An Historian's Perspective

The final presentation of the conference was given by a woman from the UC History department named Dr. Adrienne Waal.

She said that the difficulty with finding common referents to use for learning an ET language is not a new problem. In the past, when great civilizations encountered one another, they exchanged people. Commercial traders visited. Armies invaded (and were followed by administrators and commercial traders). These encounters and exchanges led to the emergence of bilingual individuals.

We do not have this option now with the ET languages. Back in the 1500s it took months to cross the Atlantic Ocean, but it was still possible for civilizations to exchange people. Native Americans were abducted to Europe. With civilizations separated by many light-years, an exchange of people can take more than a lifetime.

This will change only when somebody finds a way to travel faster than light, or more practically (given the robustness of Einstein's theory about the speed of light being a limit), somebody finds a common mode of communication which allows each side to exchange ideas based on common referents. Pictures are a possibility.

I suppose that sending a bit array with a primitive picture of our solar system and a male and female human is a start on this, but it doesn't go very far.

by Steve Geller

Nobody, on Earth or elsewhere, has invented a reliable and effective way to send a message to another cosmic civilization in such a way as to teach the target how to understand what was sent.

Dr. Waal said that SETI research needs to concentrate much more on developing a message concept under which an ET can teach himself how to understand what we send.

It may be that the only reliable way to send a civilization a message anything more elaborate than "we are here and we can hear you" is to repeat some of their own stuff back to them. They will surely recognize it, and be able to distinguish it from anything else. This idea has been used in numerous science fiction stories.

Dr. Waal then shared a comment from a colleague in the Biology Department, to the effect that if sexual reproduction really has been a major mover in the evolution of life, and similarly motivates evolution of ET life, then one "common referent" might well be sexual dimorphism. So sending the picture of a naked male and female human might really have been the right place to start. She gave a small smile.

Dr. Waal concluded her talk with these remarks:

> I notice a strong tendency among ET researchers to assume that any ET technology will consist of the same kit that we here on Earth have been able to work out. There's an unstated assumption that the only engineering solutions that exist are those that we ourselves have found.
>
> We can't simply assume that all ETs have a technology kit that overlaps ours to any extent.
>
> Engineering solutions are strongly driven by environmental needs. The Romans had good carts, but didn't use wheels with spokes, or seamless metal rims, until they encountered the Celts. The Incas didn't use wheels at all; just llamas. Some of their steep mountain roads looked like flights of stairs, because they were used as stairs, which the llamas could climb. Nobody in the entire Western Hemisphere developed gunpowder until the Europeans arrived and began blasting them with it. Paper didn't really get used in Europe before the printing press.
>
> As soon as the civilizations came in contact, wheels, gunpowder and paper were exchanged along with corn, tobacco and printing. Would this have happened without direct contact between beings? Could such technology exchanges happen at all when the simple exchange of informational messages takes many years, as it does between stars?

by Steve Geller

In 1400, Europe and the Americans were sealed off from each other Now, we and ET have been sealed off from each other and the seal is about to be broken. Is the Cygni Announcer desperately trying to tell us about some great technology, but he can't get us to understand him?

There has been some convergence. We and the ETs both discovered music, and we find that we can share it. Three cosmic civilizations developed the technology of packet telemetry. For some engineering problems, everyone did end up with the same solution.

Was Earth the first to discover radio? Is Earth the first civilization to discover the SSE?

Beyond this, there is the tendency to assume that the ETs themselves have to have at least approximately the same physical build as ours, with eyes, ears, limbs -- and fingers for manipulating things like a pen and a computer keyboard. But for all we know, the HD85 ETs may be doing their space research without fingers, comfortably relying on something we never thought of.

All ETs evolved within in an environment totally disconnected from ours. The ETs may not be able to see images the way we do. They might not even have a brain like we do.

We humans may yet turn out to be very much unique, so unique that we can communicate effectively only among ourselves, as seems to be the case among the whales and the birds.

That was a very impressive talk, but I have to admit that I left the conference in a state of confusion.

How different are we Earthlings from the various ETs? Does a shared level of technical accomplishment imply a convergence in technology? We sure don't seem to converge much on language.

It was a very good conference. Everyone got a lot out of it. The press was well-represented. I did my part to promote accurate journalism. A lot of what you just read made it into the Associated Press report. For many of the reporters, I had become the go-to authority on ET messages. Well, I and my fellow SETI Café patrons tried to do our best.

On the way out from the conference meeting, Felix introduced me to several people, telling them about the reports I had been writing and about hanging out at the SETI Café.

One guy asked me if I thought we were actually in contact with extraterrestrial civilizations. I said I thought some of the messages extracted by the SSE from radio telescope data were really sent by ETs, but we had already seen how easy it is for pranksters to fool us. I reminded this guy that during the Tau Morse fiasco, the pranksters had been people working at the radio telescope.

Another person asked if I had ever channeled an ET by listening to messages from the stars. I said I had not, but I knew at least one person who might have. If they wanted to talk about channeling ET, they should hang out at the SETI Café.

Finally, a guy asked me how it was possible to deliver extraterrestrial food to the SETI Café if the stars where it was being produced are many light-years distant. I told him that we writers can make fiction travel faster than light.

by Steve Geller

Shamisen Check

One of the people in the linguistics group had an elderly Japanese friend who played the shamisen, the twangy 3-stringed instrument some people thought they heard in the music from Epsilon Eridani. The linguist invited this shamisen player into the linguistics lab and had him listen to the Epsilon Eridani music. The musician said that our music did sound something like a shamisen, but he pointed out that the notes were different. He had brought his instrument with him, and he played a couple short pieces. His shamisen did sound a little different from the one in the Epsiloni music.

Just for fun, the shamisen player listened intently to one selection of Epsiloni music. The staff played it for him several times. He then tried to duplicate it on his shamisen. At first, it didn't work at all, then he changed his tuning a little and produced something quite close to the original. But something was still not right. He suggested that the strings on "our" instrument might be made of the wrong stuff.

The shamisen player was a very nice old Japanese gentleman. The linguistics staff suspected that he never understood just where the Epsiloni shamisen music had come from.

After that, from time-to-time, Cosmo played some shamisen music on the Café speakers. He played selections from Japanese artists and from Epsilon Eridani. Most of the time, people didn't notice the difference.

Still Trying to Understand an ET Language

The linguistics group finally developed a standard transcription for the Cygni and Epsiloni voices. They made the same transcription scheme work for both. For 61 Cygni, they used "$" for the [peep!] which is so startling in the Announcer's language. I guess they thought the Epsilonis wouldn't be talking dollar-denominated finance.

Having established their consistent transcription scheme, the linguists were able to compile written word lists. They then began to look for common prefixes, case endings and tense markers. They were doing all this without having any idea what information was being communicated, or what any of the words meant, so their description of the language remained rather ragged.

The portion of the Epsilon Eridani broadcast that was in Spanish remained a source of much frustration. The linguists were hoping to use the Spanish as a key to the Epsiloni language, but if there is any "Rosetta Stone" passage (the same thing in both Spanish and Epsiloni), it was not obvious. All they had was the example Pablo and I had found of an Epsiloni quoting Spanish with a bad accent, and probably having no idea of the meaning. The linguists were coming to the conclusion that the Epsilonis were just as mystified by Spanish as we were mystified by Epsiloni, which was possible since the Epsilonis didn't have any common referents either.

by Steve Geller

The most success in language understanding, such as it was, came from Starman's zombies, who were working as an informal group, based at the SETI Café. They listened to music and claimed to translate Cygni voice. It was hard for the linguists to say that the zombies are wrong, because the linguists were not sure what was right. Starman told me that the zombies had tried a new technique. On one audio channel, they listened to the voice and on the other channel, simultaneously listened to the corresponding music. I tried a little of that, using Starman's computer, but didn't get any improvement in my understanding of the Cygni voice.

I called Father Paul Beni, to see if he had anything to report from listening to the "sermons" by the Cygni Announcer. He said he had given up on that project, because he couldn't understand enough of what was said. I mentioned the voice from Epsilon Eridani, and he said he would try listening to that. He sounded a little annoyed with me.

Then he told me that he had been listening to MV. Somebody had told him about it.

"I think this music may have a divine origin," he said. "It may be God's way of giving us a vision of the afterlife, of Heaven."

I told him I was pleased to hear that, and suggested that he be careful.

Felix Group Listens to Red Dwarfs

About this time, the Astronomy department held one of their regular SETI seminars. Among other topics was the success of the SSE in finding ET messages where there had been none. Several attendees suggested that astronomers re-visit some stars, looking for ET messages that they might have missed before the advent of the SSE. Some university discretionary funds were found to fund the re-visits.

To start with, Felix and a group of his fellow astronomy graduate students began a project of listening for messages coming from "M-class Dwarf" stars.

The SETI search for ET messages had tended to focus on stars similar to the Sun, that is, mid-sized yellow or orange stars. The search did not pay much attention to the smaller stars for which the habitable zone might be too narrow or too close to the star for planets with ETs to form there.

Yellow stars like our Sun are fairly easy to find, but the most common type of star in the Sun's stellar neighborhood, and in the universe generally, is the red dwarf. Although called "red," the surface temperature of such a star would actually give it an orange hue when viewed from nearby.

Red dwarf stars tend to flare. This may have something to do with the fact that such stars have just barely begun nuclear energy production and have not stabilized the process. They put out sudden bursts of energy which might fry an orbiting planetary civilization. Our Sun puts out flares too; astronomers usually call them coronal mass ejections. The flares from red dwarfs can be much more energetic than solar flares.

by Steve Geller

A flare observed on the nearby star "II Pegasi," was about a hundred million times more energetic than the typical solar flare. The Pegasi flare star is only slightly less massive than the Sun. Were a comparable event to occur on the Sun, it would result in a mass extinction of life on Earth due to the outpouring of lethal X-rays. Fortunately, our Sun is a stable star that doesn't produce such powerful flares (so-far).

One day, Felix came to the SETI Café to give me a report on what his red dwarf study group had turned up. He began by telling me about their first target.

"The star is called Gliese 581. It's also known as HO Librae – it's in the constellation Libra. We've been calling it G58 for short.

"G58 is a red dwarf star of spectral type M3V, located about 20 light-years distant. Its estimated mass is about a third that of the Sun. Observations suggest that this little star may have as many as six planets, designated Gliese 581 e, b, c, d, f, g, going from inner to outer. (the letters were assigned in order of discovery, not distance from star) . The g planet is thought to be within the star's habitable zone, although that's not real clear. Gliese 581b is large and very close in; it might be a hot Neptune."

I interrupted this very detailed recitation. "Felix, is it possible that you chose this star for your project because it is on the SETI Café menu as having inspired "Red Star Beans, Beets and Onions?" I pointed to the item on the menu.

Felix grinned cheerfully. "Why sure," he replied. "I was sitting here one day, wondering what would be a good M-class dwarf to study and there was one right in front of me. I think Dr. Stern put that star on the menu."

"So what exactly did you turn up?" I encouraged him.

Felix continued his tale. "We got some radio telescope time at Stanford and tried some likely frequencies. We got a signal on two frequencies. One was very poor; it might have been voice, but it was too broken up to tell. The other frequency had a solid signal; we got a burst of about 15 minutes."

I asked, "Can you tell which planet the signal is coming from?"

Felix shrugged. "My guess is we're hearing the 'g' planet, which has an orbital period of 37 days. Anyway, it seems likely we heard a radio station which was briefly beamed out toward Earth through a temporary hole in g's ionosphere.

"We still don't understand why the signal is so strong. A distance of 20 light-years should wash it out if there's nothing to lens it. One of our group thinks the signal is being deliberately beamed up and out through the planet's atmosphere, perhaps using an ET satellite. Conceivably, the aliens are actually probing their sky for ET civilizations. If so, they've now found one — us Earthlings!"

"Did you say what's on the signal?" I prodded. "Is it music or voice?" "It's both," Felix replied. "It's actually kind of strange. Some of it might be music. It's kind of a whiney humming. Other parts are definitely a voice. We've sent recordings to the linguistics group. Here, you can listen to some of it." Felix looked like he knew a secret.

by Steve Geller

Felix activated his laptop and then handed me his earphones. "There's some of the music first, then a complete burst of the voice."

I heard a thin eerie wailing, which reminded me of a Theremin. The note varied and the volume rose and fell. The music wasn't very interesting. I could believe I was hearing a single-sideband radio station being tuned -- "wee-ooo-eee-ooo."

Then, abruptly, the voice cut in. It was deep and authoritative. It sounded like a dictator's political speech. The language didn't sound at all like the other ET voices that had been received. I didn't understand any of it.

"One of my friends says the voice is speaking in Klingon," Felix said, with a smirk.

"Could be," I replied. "The language is kind of harsh, with a lot of 'sh' sounds in it. Maybe G58 is 'Kronos,' the Klingon home world in Star Trek?"

"Yeah, right," replied Felix, chuckling, "that sure would surprise the producers of Star Trek, wouldn't it? One of my friends sent a recording of our Klingon speaker to a Trekkie fanatic he knows."

"We studied several other red dwarfs too. One was Proxima Centauri. As you know, that's the actual nearest star to us. In the Alpha Centauri system, there are two sun-like stars in close orbit, Alpha Centauri A and B. The C star, called Proxima because it's nearest Earth, is a red dwarf orbiting 15,000 AU away from the center of mass of A-B. That far out, a quarter of a light-year, Proxima may not actually be in orbit, but just happened to turn up where it is as it wandered through the galaxy.

"All we got was some flare noise. If Proxima has any close planet, the flares would probably be very dangerous for life there.

"Recently, somebody discovered a planet orbiting Alpha Centauri B. Not verified yet, but it might be Earth-sized.

"Another red dwarf we looked at was Groombridge 34. That's a binary system, 11.7 light-years distant in the constellation Andromeda. It produced no ET signals.

"Another was Gliese 623. That's another binary; both components are red dwarfs, 25 light-years distant in the constellation Hercules. There was a hint of a signal, but it was not strong enough, or stable enough, for us to be sure we had anything. I think some flare noises can make the SSE do a kind of twitch. The SSE didn't really pick out any ET radio signal there.

by Steve Geller

"Barnard's Star is a single red dwarf, only 6 light-years distant. Barnard's Star has approximately 14% of the Sun's mass; it has a radius 15% to 20% of that of the Sun. Barnard's Star is famous for having the highest 'proper motion' of any star. This means that the star's rate of movement within the galaxy results in the largest shift on Earth's sky, per year, of any star, even more than that of "Piazzi's Flying Star" 61 Cygni. Barnard's seems to be a very old star. It may have planets, but this hasn't been demonstrated. In the early 1970's the astronomer Peter Van de Kamp claimed that there was a gas giant planet (maybe more than one) in orbit around Barnard's Star. This claim was refuted when measurement errors were discovered. When the SSE processed the radio emission from Barnard's Star, it saw no ET signals, just some flare noise from the star itself."

"So you essentially got a null result?" I ventured.

Felix replied, "Not totally. We did get some music from G58, even if that Klingon voice turns out to be a spoof. "I claim that we showed that nearly all of the nearby red dwarf stars do not harbor any ET civilizations – unless they're very quiet and reclusive. If anyone is there, they've got to enjoy being frequently zapped by flares." He smiled and shrugged. "Anyway, it was a good exercise for us astronomy graduate students. I think we'll write a paper covering what I just summarized for you."

What is the Gliese Catalogue?

We had been hearing about stars named "Gliese" something. So I checked Wikipedia. Here's a summary:

The German astronomer Wilhelm Gliese (GLEE-zuh) began compiling a catalog of nearby stars in 1957. His first catalog listed almost 1000 stars within 83 light-years of the Sun.

Gliese published an extension to the second edition of the catalogue in 1979 in collaboration with Hartmut Jahreiss. Stars in this combined catalogue usually have the designation "GJ", for example GJ 667.

The catalog has since been augmented by other astronomers, and the celestial coordinates of the stars updated to reflect the Hipparcos astrometric measurements.

The current Gliese Catalog of Nearby Stars was published in 2010 and consists of 3,803 stars.

It looks like we have plenty of neighbor stars.

by Steve Geller

Brown Dwarfs

There was one other category of dwarf star that Felix and his group tried. This is the "brown dwarf," the coldest spectral class of stars.

Like regular stars, brown dwarfs start their lives as contracting balls of gas. In stars like the Sun or a red dwarf, the energy released by this contraction makes the core hotter and hotter until nuclear fusion starts. Brown dwarfs get hot inside, but the nuclear fusion never starts. These objects are not quite stars. The larger ones might do occasional ragged bursts of nuclear fusion, but most do no fusion at all. They generate all their heat from gravitation, from compression, and they radiate entirely in the infrared.

The first brown dwarf they studied is called WISE 1541-2250 (discovered in 2011 by the Wide-field Infrared Survey Explorer). Felix called it W15 for short. W15 is in the constellation Lyra, and not far in the sky from the bright star Vega.

W15 has a mass somewhere between 8 and 12 times Jupiter. It has a surface temperature about 350 K vs the Sun's 5700 K. (Boiling water is 373 K) As stars go, W15 is a cold body. It's not certain that a brown dwarf has a solid surface; it might be gassy like Jupiter. Perhaps with shoes on, you could walk around on the warm surface of W15, but it would be hard slogging, because, depending on the dwarf's radius, you would weigh somewhere between 5 and 20 times as much as on Earth.

The distance of W15 is uncertain, between 9 and 40 light-years. If this star has any planets, they are very probably too cold to have evolved intelligent life, and I guess nothing had evolved on the star's warm surface to the point where it constructed radio transmitters. There were no ET signals from the region of W15.

Felix's group also looked at one close neighbor, a pair of dim brown dwarf stars 6.5 light years distant. They are called WISE 1049-5319, they are the third-closest star system to us, after the Alpha Centauri triple star about 4.3 light years away, and Barnard's star about six light years distant.

Felix's red dwarf group also looked at a newly-found brown dwarf named WISE J085510.83-071442.5. It may be the coldest "star" on record. It has a chilly temperature between −54° and 9° Fahrenheit (−48° to −13° Celsius). Previous record-holders for coldest brown dwarfs, also found by WISE, were about room temperature.

WISE J085510.83-071442.5 is estimated to be three to 10 times the mass of Jupiter. With such a low mass, it could have been a gas giant planet similar to Jupiter that was ejected from its star system. But scientists estimate it is probably a brown dwarf rather than a planet since brown dwarfs are known to be fairly common. If so, it is one of the least massive brown dwarfs known.

It's faintly possible that the surface of a brown dwarf itself could have life, in warm oceans heated from below. This possibility makes it urgent to find out more about oceans on Jupiter's moon Europa and Saturn's moon Enceladus.

I guess brown dwarfs are much too cold for an ET civilization to develop. Felix found no signals at all coming from these stars.

by Steve Geller

Felix's group also processed a recently-discovered brown dwarf known as Teegarden's Star. This star is very dim, but quite close, 12 light-years distant in the constellation Aries. Teegarden's Star was found to have a very large proper motion, Like Barnard's Star. It was discovered in 2003 by re-examining asteroid tracking data that had been collected years earlier. The star is named after the discovery team leader, Bonnard Teegarden, a NASA astrophysicist. As with the WISE objects, the SSE could get no ET signals from Teegarden's Star.

Felix said that astronomers have long thought that there exist many undiscovered dwarf stars within 20 light years of Earth, as stellar population surveys show the count of known nearby dwarf stars to be lower than otherwise expected and these stars are dim and easily overlooked.

This work fascinated me. There are so many different kinds of stars out there. But so-far, it doesn't seem likely that any red or brown dwarf star will turn out to be home to an ET.

Another Fraudster Found

To hardly anyone's surprise, the Klingon voice from G58 turned out to be a joke.

This was exposed when some Trekkies heard the gruff G58 voice and said it really was the Klingon language from the fictional civilization in Star Trek. That language was deliberately designed by Marc Okrand to be "alien."

What happened was that somebody had managed to get around our local anti-fraud protections. They took some real Klingon speech (recorded from a Star Trek episode) added some Theremin music and combined it all to modulate a signal which was made to look like it came from G58. It was not clear where and how the fraudulent signal had been introduced. Yow; these fraudsters put out so much effort just to fool people.

There had been signals from G58 at two frequencies. One was poor and broken up. The other was the one that had the Klingon; that's why the signal seemed too strong.

by Steve Geller

A Signal from Sirius?

One night, on her way home, Dr. Stern had looked up at the night sky and had seen the familiar human-shape of Orion the hunter. She knew that the bright stars in that constellation were very far away. Red Betelgeuse, at the upper left, had provided some interesting artificial ET Music for a while, but at 640 light-years it's way too far away and nobody has found any packet messages or real ET music. Her eyes swept down off the body of the hunter and off to the left, into the constellation Canis Major, the hunter's big dog. And there was Sirius, blue-white, bright and only about 9 light-years distant. We ought to get something from Sirius, she thought.

During the next month, she put in a proposal for some of the re-visit money that Felix had been using to look at red dwarfs. Her proposal was to do a quick SETI survey of a few well-known bright stars, running the results through the SSE. Some of those stars had not been looked at for ET messages for a long time, if ever.

She got the money and arranged for some radio telescope time, first pointing at Sirius. Several frequencies were tried, statistically enhanced and run through versions of the same post-processing software that had found the packets from HD85 and HD28 and had brought in the various sources of voice and music.

They got a lot of signals, some from big Sirius A, some from the white dwarf companion Sirius B. There was nothing like telemetry packets or FM radio — not even Sirius Morse Code. It was kind of a disappointment, because the star Sirius looms so large in Earth's mythology.

Checking on Canopus

After getting nothing from Sirius, Dr. Stern looked at Canopus. This is a bright yellowish-white star in the southern hemisphere. It is big, 65 times the size of the Sun, and hot, with a 7350K surface temperature.

Canopus is far south in the sky. It never rises in mid- or far-northern latitudes; in theory the northern limit of its visibility is latitude 37°18' north. This is almost exactly the latitude of Lick Observatory on Mt. Hamilton, California, near San Jose. Astronomers on the mountain can see Canopus, because of the effects of elevation and atmospheric refraction, which add another degree to its apparent altitude.

Bright Canopus is regularly used as a reference star to orient a spacecraft. It is so far away that it has almost no proper motion; it stays put. NASA literature often mentions use of a "Canopus Star Tracker."

It is a supergiant. For Canopus to appear the same size as our Sun, Earth would have to be moved from 1 AU out to 65 AU (about the distance of Pluto).

Canopus is a strong source of X-rays, which are probably produced by its corona, magnetically heated to around 15 million degrees K.

So-far, no planets have been detected orbiting Canopus. The X-rays would probably sterilize them anyway. Canopus is pretty far away – more than 300 light-years. Not too surprisingly, the SSE did not detect any ET signals from it.

by Steve Geller

Dismissing Deneb

An attempt to get messages from any ETs at Deneb was a non-starter.

Deneb is the brightest star in the northern constellation Cygnus the Swan. It's a blue-white supergiant. It is among the most luminous stars known, with an estimated luminosity nearly 200,000 times that of our Sun. But Deneb is very far away; its distance is rather uncertain, even with the results from Hipparcos. The best guess is at least 1,340 light-years and maybe much farther. This was too far even for the mighty SSE, so no further work was done on Deneb. It was probably too hot for life anyway.

Vetting Vega

Vega, the brightest star in the northern hemisphere constellation Lyra, appears to have a circumstellar disk of dust, probably the result of collisions between objects in Vega's version of the solar system's Kuiper belt. Vega is twice as massive as the Sun and is 25 light-years distant. Irregularities in Vega's dust disk suggest the presence of at least one planet, likely to be about the size of Jupiter.

Vega has an unusually low abundance of "metals," the elements with a higher atomic number than that of Helium. This doesn't bode well for Earth-like planets. Vega also may vary slightly in magnitude in a periodic manner.

The SSE processed several signals from Vega, but none of them appeared to have intelligent content. If there's an Earth-like planet orbiting Vega, it may be in the first stages of formation from that dust disk.

Follow-up with Fomalhaut

Dr. Stern tried one additional bright star. Fomalhaut, in the southern hemisphere, is 25 light-years distant. It is visible in the Northern Hemisphere in autumn, very low on the southern horizon, in a region where there are few bright stars. Its mass is twice that of the Sun. Its radius is about 1.8 times the Sun and its surface temperature is cool, only 875K.

Fomalhaut has a huge debris disk and a dust ring surrounding it, 25 AU wide, left over from its formation. It is a young star, only between 100 and 300 million years old. Planet formation may be taking place right now. In fact, it's possible for a big telescope to actually see what looks like a planet embedded in the debris disk. However, the planet might be just a particularly thick blob of debris — which could condense into a planet later on.

The SSE pulled out a variety of miscellaneous signals from Fomalhaut. It was not clear whether they came from the star, the debris disk or the possible planet. Fomalhaut sent no ET music or voice.

Felix wanted to try some big stars, like Betelgeuse and Antares. He also wanted to study the bright blue stars that make up the Pleiades star cluster. But Dr. Stern pointed out that these big hot stars are poor candidates for planets with ETs in residence. Besides, they'd about run out of grant money.

Well, as with the red dwarfs, they had collected a lot of interesting data and had the makings of a paper to publish.

Images from an ET

Things had been quiet for a while.

The SSE-enhanced SETI surveys of the familiar bright stars and of nearby red dwarfs had not delivered any great revelations. That re-visit project was now at an end.

The familiar ET message sources continued to deliver the same stuff — mysterious music and incomprehensible voice.

Then finally, our oldest source showed us an image.

During the noon hour one day, Felix came into the SETI Café looking for me. He came over to my table. He asked me to finish eating my Epsiloni lunch and then come with him to the campus.

"There's something going on at the SSE lab that you ought to check out." Felix was rather mysterious; he wouldn't tell me any details.

We walked along, crossing Oxford Street from west to east and continuing uphill into the UC campus. After we had gotten well away from downtown Berkeley and the SETI Café, Felix finally confided his information and explained why he had been so secretive.

"You see, I wanted to avoid attracting the attention of Starman or any of the other excitable people. The SSE lab has captured some pictures from HD85."

Well, that sure got my attention. "Pictures of what?" I demanded.

by Steve Geller

Felix explained. "The SSE lab found signal data on a new frequency coming from HD85. It appears to be a group of packets which form an array. They actually first encountered this array a week ago. Since then, they have spent a great deal of time analyzing what they had. Now they think they have determined that the array is 720 by 540 pixels. Interestingly, that is the same 4:3 aspect ratio as traditional TV here on Earth. The SSE group has made some guesses about how to read brightness from the pixel elements in the packets. They have produced a monochrome image. There's probably color information there too, but they haven't yet figured how that is encoded."

"Wow!" I exclaimed. "That's amazing. I take it that this array of packets could not have come from Earth's digital TV or satellite telemetry. Right?"

Felix grimaced. "Well, there's always the possibility that, once again, somebody salted SETI data, but these days the SSE lab has very stringent fraud protection procedures in place. The staff members are quite sure the array really did come from HD85. Some of them think the packets are images being telemetered back from the same spacecraft that HD85 has been sending those other packets to, the ones with all the numbers."

Felix and I soon arrived at the building housing the SSE lab. When we entered the lab, we saw a small crowd gathered around the computer monitor at one of the analysis workstations. They were all young people, grinning with excitement.

"I think they are looking at the image now," said Felix. We joined the group.

The image on the monitor looked like a picture of one of the moons of Saturn taken by the Cassini mission. The limb of the moon, planet or whatever it was, made a large arc through the center of the image, separating pale gray moon from black space background. The moon's surface was covered with craters.

One of the young people greeted us. "Hi Felix. You know Agnes and Paula from the planetary imaging group, don't you? They say this picture definitely was not taken in our solar system. The image looks kind of like Mercury, our Moon or one of the smaller moons of the outer planets, but it isn't any of them. Right, Paula?"

The dark-haired young woman addressed as Paula nodded agreement. She smiled brightly and said, "I bet this image will be featured on HD85 commercial TV news, back at their home planet. It looks like the HD85 Voyager Mission has scored a big success. That must be one of their outer planets, or a moon of something out there."

Felix turned to one of the SSE staff and said, "Nathan, you know she might be right. I wish we could tune in to the leakage from HD85 broadcast TV." Nathan shrugged and replied, "Right, sure, well, we'll keep trying."

"Is this the only image?" Felix asked Nathan.

"No, we pulled together a few more. Want to see them?" Felix nodded. Nathan went to the computer and started working. A new image appeared. It looked like the same moon, this time filling the frame.

by Steve Geller

"Here's another. See what you make of this." The new image was mostly black, but there were numerous bright specks in it. Nathan turned to a co-worker and asked "Ezra, have you finished the pattern matching?"

"Yeah," was the reply. "I've been waiting to tell you. Maybe this picture should be on <u>our</u> TV evening news." Ezra gave a broad grin.

"What do you mean?" asked Nathan. He turned to the gathered crowd. "Oh, folks, this is a picture of a star field probably taken from the HD85 spacecraft. Ezra here has been working on matching the star pattern to see where it is pointing. What are we looking at?"

Ezra used the pen he had in his hand to point to one of the bright stars in the image. "I think that's our Sun."

This startled everyone. After the hubbub died down, we got a detailed explanation. Ezra had generated simulated star fields – the neighboring stars seen from HD85, looking out in various directions. He used these star fields to match with the HD85 star field image. He got a match for an image looking from HD85 toward Earth, and was able to identify our Sun as a small stellar point in the HD85 image. Somehow, we had managed to intercept the transmission of images from their Voyager-like spacecraft, probably directed toward the HD85 home planet – and toward us!

This was awesome. It was the first picture to be received in an ET message, and it showed our own Sun as part of the star background seen from a place 36 light-years distant from Earth.

The point Paula had been making was that the ETs would publicize their own space mission success, featuring these pictures on their news programs. Knowing this, the SSE team may possibly be able to find the same image pattern in a leakage signal, which might tell us how to decode HD85 local TV.

Felix got the SSE lab to send a copy of the image with our Sun and another image with the crater-strewn moon to the SETI Café website. He also printed some hard-copies.

We carried the image printouts back to the SETI Café and posted them as part of the collection on the walls. Felix spent the next hour explaining the pictures, again and again, to a string of arriving Café customers and a few reporters.

The HD85 star field picture with the Sun appeared in the papers the next day. It probably was on TV too, but I missed it.

by Steve Geller

Serious Speculation on the MV

None of the ET message conferences had much to say about MV – the mysterious Music from the Void.

We got new MV music every so often. No radio station ever laid claim to producing this music. None of the local musicians recognized the composer, although they recognized similarities with Earth musical instruments.

I, Starman and several others have tried to get visions by listening to the latest MV music. I got a vision of the Epsilon Eridani system seen from space. It strongly resembled the pictures on the Web, so I don't think the music was conveying any new data to me – just prodding my memories.

One Friday, at the SETI Café, an informal conference was held to discuss MV. The meeting wasn't called by anyone. All the participants just happened to be there. Expertise was contributed, in roughly descending order of competence, by Dr. Stern, Felix, Cosmo, Abner, Starman and me.

Since Abner and Cosmo had never designated an "inspired by" dish for MV, each of us ordered whatever was our favorite. I stuck with my Epsiloni salad; Dr. Stern chose Tau Tofu. After preparing everyone's order, Abner mixed Cosmo and himself some Jupiter Red Spot Yogurt and sat down with us. It was 8:30 p.m., getting near closing time.

Dr. Stern summarized the issue before us.

"The easiest thing to do is just write off MV as music from Earth which somehow appears to be coming from the sky. Or, we assume that someone salted the radio telescope data." Dr. Stern gave a sigh. "Well, we've seen it happen all too often, haven't we? But, we now think we have good data security at the radio telescopes and the SSE lab, so it's unlikely the MV music is being deliberately inserted into the recorded signal. By the way, it is just the one frequency which is being modulated with MV, and the signal is coming from a very definite place in the sky, in Ursa Major.

"There's no varying Doppler, either than what results from Earth's motions. MV seems to be standing still out there, which really bothers me as an astronomer – I expect everything to be in orbit about something!

"And of course there's that pesky smudge of a galaxy, which makes it look like MV was transmitted over 8 million years ago.

"All these indications could be mistaken, due to some physics effect we still haven't realized that we must account for.

"And while the music is pleasing to an Earthling's ear, the musical notes and other conventions in MV are not quite the same as those used by today's composers of instrumental music. But they're close."

Starman spoke up: "Doesn't all that tell us that MV is definitely extraterrestrial? It doesn't have to be transmitted from a planet. We may well have picked up the signal of an ET who doesn't live on a planet."

Cosmo made a suggestion. "Maybe MV is leaking through from elsewhere (or elsewhen), using something like a cosmic wormhole. It could even be coming from a parallel universe."

by Steve Geller

"That makes good sense to me," said Abner. "This music is so beautiful, it could have been played by spirits out in space, not living beings at all."

I had to comment on that. "I remember the remarks by the historian who spoke at the end of the last conference. She said we shouldn't assume that we and ET have a shared technology kit. We might be hearing from an ET who is very much different from us, who has worked out very different solutions to engineering problems.

"And Abner might be right that the source is something like spirits, something immaterial like a wobbling cosmic magnetic field."

Dr. Stern tried to pull us a little closer to known physics. "Keep a grip on Occam's Razor. We shouldn't bring in too many fantastic ideas. This reminds me of the debate over Dark Matter. We know it's there; we can see its effects. It needs to be explained. Physicists have suggested unknown particles and unknown aspects of gravitation which only show up at large scales. Maybe Dark Matter is just heaps of Higgs Bosons." She laughed, then said "Leakage of MV from a parallel universe is possible, but I think Occam's Razor might shave that one away."

Cosmo, Abner, Pablo and Doris were ready to close the Café, so Felix made a quick summary: "We may just have to leave MV a mystery. People used to do that for the origin of rainfall, eclipses and earthquakes. Today, as clever technologists who believe in science, we don't feel comfortable attributing stuff we can't understand to the mysterious actions of the gods and the spirits, but what's left?"

"Yeah," said Starman with a dreamy smile. "Maybe it's Angels."

by Steve Geller

Finally, A Possible Source for MV

A few days later, Felix brought us what may be at least a partial solution to the MV mystery.

The optical telescope at Lick Observatory had once again stared hard at the MV coordinates, but this time looking in the infrared.

They found a faint star there, either a dim red dwarf or a brown dwarf.

It was definitely dim, and easy to miss; all of its light was in the infrared. There was no sign of a planet.

The MV star may be similar to W15, a brown dwarf studied earlier by Felix's group.

W15 has a surface temperature about 350 K (boiling water is 373 K) and a mass somewhere between 8 and 12 times Jupiter.

It was possible that this failed star was cool enough to have retained a surface firm enough for ETs to live on, and it was from a transmitter on that warm surface that we had been serenaded with the harmonized hum tones of Music from the Void.

But living on such a surface would be like walking around on a hot plate, under very strong gravity, because the failed star would still have a mass many times that of Jupiter.

I didn't like the warm hard surface idea. I still wanted to bet on an unseen orbiting planet, but then I remembered that there ought to be a varying Doppler in MV.

Hey, maybe there was a Doppler and it was making the tone notes rise and fall. Yeah, right.

Anyway, now it seemed quite likely that there was an actual object at the location from which MV was being received.

So we stopped talking about the MV coming from the far side of the universe. This dim star had to be close; it could not be more than 20 light-years distant. The shift in the infrared spectrum showed a steady radial velocity of 15 km/sec (approaching), which is about right for a nearby star.

This MV denouement was something of a disappointment, but on the other hand, it encouraged us to believe that even some failed stars could send us ET messages.

We had now received many messages from the stars. SETI had succeeded, far more than anyone had expected.

End of the Story

You can order more paperback and an ebook version by going to amazon.com

For other writings by Steve Geller, go to his website:
http://berkeleybus.mysite.com/

Steve Geller may be contacted at sgeller6@gmail.com

Bibliography

Some books which helped in the writing of
"When the Stars Began to Speak"

Web page on the **History of SETI**
http://archive.seti.org/seti/seti-background/

Contact, by Carl Sagan (SF novel)
Pocket Books -- July 1977
ISBN: 0-671-00410-7
This is the famous story, made into a movie, about aliens sending Earth plans to build a spaceship.

Rare Earth: Why Complex Life Is Uncommon in the Universe
(nonfiction) by Peter Ward and Donald Brownlee
Copernicus Books, 2000
ISBN: 0-387-95289-6
This book deals straightforwardly with the possibility that life on Earth, and especially human life, may be quite rare in the universe. It points out the unique history and attributes of our planetary home.

Lucky Planet by Dave Waltham
(nonfiction)
Same subject as "Rare Earth"

Contact with Alien Civilizations:
Our Hopes and Fears about Encountering
Extraterrestrials (nonfiction)
by Michael A G Michaud
Springer Science+Business Media, 2007
ISBN: 0-387-28598-9
Very good coverage of the technical issues
involved in SETI, and the prospects for its success.

The Listeners, by James E Gunn
(SF novel)
BenBella Books edition - January 2004
ISBN: 1-932-100-12-1
Story about the possible future of SETI, and
what might happen with success.

SETI Breakthrough, by Tony Manera
(SF short story)
Amazon Kindle edition, November 2011
Another take on what happens with SETI success.

Omnilingual, by H. Beam Piper
(SF novel)
Amazon Kindle edition, October 2011
ISBN: 978-1428073128
About an archaeological study of the (fictional)
ancient civilization on Mars, and how the Earth
scientists learned how to read the Martian language.

For the Sagan and Salpeter speculation about **Jupiter life**, go to website
www.daviddarling.info/encyclopedia/J/jupiterlife.html

A Meeting with Medusa , a short story by
Arthur C. Clarke in the
collection "The Wind From the Sun".
The story imagines Jovian aerial life-forms
like those described by Sagan and Salpeter.

Confessions of an Alien Hunter, by Seth Shostak
(nonfiction)
National Geographic 2009
ISBN: 978-4262-0433-3

First Contact
(nonfiction)
by Bob Connolly and Robin Anderson
Penguin Books, September 1988
ISBN: 978-0140074659
In the 1930's, some Australian gold prospectors
entered the mountain highlands of the island
of New Guinea, and encountered "aliens" – tribes of
humans who had never before had contact
with other humans.

www.ingramcontent.com/pod-product-compliance
Lightning Source LLC
Chambersburg PA
CBHW071414170526
45165CB00001B/273